Seaforth WORLD NAVAL REVIEW 2017

1000

Seaforth

WORLD NAVAL REVIEW

2017

Editor
CONRAD WATERS

Seaforth
PUBLISHING

Frontispiece: The futuristic new US Navy destroyer *Zumwalt* (DDG-1000) seen on trials in the North Atlantic in April 2016. Decisions to cease construction of the class at just three vessels mean that the class will be a rare sight on the world's oceans but their innovative technology is likely to have a significant impact on future US Navy designs. *(General Dynamics/Bath Iron Works)*

First published in Great Britain in 2016 by
Seaforth Publishing
An imprint of Pen & Sword Books Ltd
47 Church Street, Barnsley
S Yorkshire S70 2AS

www.seaforthpublishing.com
Email info@seaforthpublishing.com

British Library Cataloguing in Publication Data
A CIP data record for this book is available from the British Library

ISBN 978-1-4738-9275-0

Typeset and designed by Stephen Dent
Printed in China by 1010 Printing International Ltd.

CONTENTS

Section 1: Overview

1.0 **Introduction** 6

Section 2: World Fleet Reviews

2.1 **Regional Review – North and South America** 12

2.2 **Regional Review – Asia and The Pacific** 26

2.2A **Fleet Review – The Royal Malaysian Navy: A Difficult Balancing Act Ahead** 42
Mrityunjoy Mazumdar examines the navy of a country divided by 500 miles of water

2.3 **Regional Review – The Indian Ocean and Africa** 54

2.4 **Regional Review – Europe and Russia** 64

2.4A **Fleet Review – The Royal Danish Navy: From the Baltic to the Horn of Africa – And Back Again?** 82
Søren Nørby highlights how Denmark's maritime horizons have shifted since the Cold War's end

2.4B **Fleet Review – The Royal Netherlands Navy: Recent Developments and Current Status** 92
Theodore Hughes-Riley assesses the recent direction one of the world's most innovative naval forces

Section 3: Significant Ships

3.1 ***Akizuki* Class Destroyers** 104
Tomohiko Tada provides a detailed description of Japan's latest destroyer class

3.2 ***Dokdo* Class Amphibious Assault Ships: South Korea's LPX-1 design** 120
Guy Toremans explains the importance of the class to the Republic of Korea's broader naval ambitions

3.3 ***Zumwalt* (DDG-1000): Past and Future Tense** 136
Edward Feege and Scott Truver analyse the importance of the US Navy's radical surface combatant

Section 4: Technological Reviews

4.1 **World Naval Aviation** 156
David Hobbs undertakes his annual overview of recent developments in maritime air power.

4.2 **Shipboard Anti-Ship Missiles: Some Considerations** 174
Norman Friedman looks at a weapon that is generating renewed interest in the US Navy

4.3 **SMERAS: Submarine Escape, Rescue, Abandonment & Survivability** 182
Jan Ziolo expounds the need for preparation and speed when assisting a submarine in distress

Contributors 192

Note on Tables: Tables are provided to give a broad indication of fleet sizes and other key information but should be regarded only as a general guide. For example, many published sources differ significantly on the principal particulars of ships, whilst even governmental information can be subject to contradiction. In general terms, the data contained in these tables is based on official information updated as of June 2016, supplemented by reference to a wide range of secondary and corporate sources, such as shipbuilder websites.

1 OVERVIEW

INTRODUCTION

'Against danger, it is best to be prepared' wrote the Ancient Greek storyteller Aesop in his fable of *The Wild Boar and the Fox*. The moral of the story seems particularly relevant at the present time as risks from both state-related tensions and non-state-related violence continue to dominate the newspaper headlines. Indeed, there have been some indications over the past twelve months that governments worldwide are taking heed of the danger. Notably, defence reviews in countries as far apart as Australia and the United Kingdom have committed to bolstering spending to fund both refreshed and new capabilities.[1]

An examination of the outcomes of the Australian and British defence reviews is instructive in terms of both their similarities and their differences. Interestingly, both commit to spending in the order of two per cent of national wealth – as measured by gross domestic product (GDP) – on national security. However, the Australian figure represents fulfilment of a previous pledge to increase spending that has already risen by some thirty per cent over the last decade, whilst the United Kingdom's commitment provides a floor to a long period of steady decline. Both reviews also make particular reference to the need to support an international rules-based order and the dangers posed by terrorism, as well as the rapidly expanding importance of cyber security. Expanding investment in these capabilities is a clear priority in the British review but the Australian white paper is more nuanced. With one eye to China's expansionist policies, much of Australia's planned defence investment programme is meant to facilitate conventional operations against a state opponent, primarily in the Indo-Pacific region.[2] One consequence of this is that the Australian review places much greater emphasis on expanding 'hard' power naval capabilities.

The trend towards increasing defence spending is also reflected in Table 1.0.1, which is based on the Stockholm International Peace Research Institute's (SIPRI's) annual review of world military spending.[3] Estimated global defence spending grew slightly to US$1,676bn in 2015: the first time since 2011 that there has been a real-terms increase. Unsurprisingly, expenditure in the regional hotspots of Eastern Europe, Asia and – to the extent there was data available – the Middle East showed the greatest growth. However, there were also signs that recent declines in North America and Western Europe are coming to an end as the security environment becomes more complex. What the future holds is more difficult to discern. The share of national wealth expended on the military has arguably reached unsustainably high levels in a number of countries. This is particularly so for those, such as Saudi Arabia and Russia, that have economies heavily exposed to the decline in world oil prices. As a result, both countries – currently third and fourth in the list of big spenders – are expected to see reduced defence budgets in 2016.

Table 1.0.1: COUNTRIES WITH HIGH NATIONAL DEFENCE EXPENDITURES – 2015

RANK	COUNTRY	TOTAL: US$	SHARE OF GDP: %	CHANGE 2006–15
1 (1)	United States	596bn	3.3%	-3.9%
2 (2)	China	215bn	1.9%	132%
3 (4)	Saudi Arabia	87.2bn	13.7%	97%
4 (3)	Russian Federation	66.4bn	5.4%	91%
5 (6)	United Kingdom	55.5bn	2.0%	-7.2%
6 (7)	India	51.3bn	2.3%	43%
7 (5)	France	50.9bn	2.1%	-5.9%
8 (9)	Japan	40.9bn	1.0%	-0.5%
9 (8)	Germany	39.4bn	1.2%	2.8%
10 (10)	South Korea	36.4bn	2.6%	37%
11 (11)	Brazil	24.6bn	1.4%	38%
12 (12)	Italy	23.8bn	1.3%	-30%
13 (13)	Australia	23.6bn	1.9%	32%
14 (14)	United Arab Emirates	22.8bn	5.7%	136%
15 (15)	Israel	16.1bn	5.4%	2.6%
	World	**1,676bn**	**2.3%**	**19%**

Information from the Stockholm International Peace Research Institute (SIPRI) – https://www.sipri.org/databases/milex/ The SIPRI Military Expenditure Database contains data on 172 countries over the period 1988-2015.

Notes:
1 Spending figures are at current prices and market exchange rates.
2 Figures for China and the UAE are estimates, with the UAE figure relating to 2014. Previous figures for China have been reduced downward from previous figures contained in the SIPRI database.
3 Data on military expenditure as a share of GDP (Gross Domestic Product) relates to GDP estimates from the IMF World Economic Outlook, October 2015.
4 Change is real terms change, i.e. adjusted for local inflation.
5 Figures in brackets reflect rank in 2014, revised for latest information.

Given, particularly, Russia's difficult situation, it is easy to conclude that the recent willingness exhibited by many Western European countries to spend more on their militaries in response to Russian adventurism in Ukraine and elsewhere will soon be reversed as risk perceptions are adjusted downwards. Certainly, Russia's navy continues to pay a heavy price for its intervention in the Crimea and the Donbass, most notably through the impact of sanctions on its modernisation plans. Its French-built *Mistral* class amphibious assault ships now fly the Egyptian flag, whilst progress on important surface programmes continue to be slowed by lack of components.[4] In spite of this backdrop, it appears that many countries have undertaken a fundamental re-appraisal of the risks they face – perhaps for the first time since the end of the Cold War – and have become concerned by the result. There appears to be a broad consensus that higher levels of spending will be sustained for the rest of the decade.

One further element of uncertainty, however, arises from the United Kingdom's vote to leave the European Union that was announced on 24 June 2016, just before this annual's cut-off date. The consequences will not be felt only within the United Kingdom, including a likely revisiting of the relationship between Scotland and the rest of the Kingdom, but also more widely throughout Europe. In the words of *The New York Times*, 'For the European Union, the result is a disaster, raising questions about the direction, cohesion and future of a block built on liberal values and shared sovereignty that represents, with NATO, a vital component of Europe's post-war structure.'[5] It seems inevitable that the decision will have large implications for both British and European security structures but it is too early to make a meaningful assessment of what these might be.

The Royal Australian Navy's *Collins* class submarine *Rankin* pictured operating with one of the navy's new MH-60R Seahawk helicopters in March 2016. Australia is one of a number of countries bolstering defence spending against a more threatening international background. Its 2016 *Defence White Paper* confirmed plans to double the country's submarine flotilla as the existing boats are replaced, with a conventional derivative of DCNS' 'Barracuda' design being subsequently selected for the contract. *(Royal Australian Navy)*

FLEET REVIEWS

Given protracted implementation times for major defence programmes, recent decisions to bolster defence spending in many Western countries inevitably have negligible impact on the estimated strengths of the world's major navies outlined in Table 1.0.2. For example, work on Australia's Project SEA 1000 – the replacement of the existing *Collins* class submarines with a greatly enlarged force – commenced in 2007.[6] However, the recent white paper confirmed that it will not be until the early 2030s that the first boats begin entering service. That being said, the changing security environment will inevitably have a somewhat more immediate major impact on the 'direction of travel' on a number of fleets. This is particularly the case for many of the medium-sized European navies, which underwent a major transformation from an anti-submarine and littoral defence configuration at the end of the Cold War to a new emphasis on expeditionary stabilisation operations.[7] Having now largely completed this painful reconstruction, they face an at least partial renewal of the original Cold War threat.

Two of the European fleets that are 'victims' of this somewhat ironic turn of events are the Royal Danish Navy and the Royal Netherlands Navy, analysed in this issue by first-time *World Naval Review* contributors Søren Nørby and Theodore Hughes-Riley. Both countries are historic maritime powers that retain significant economic interests in the safety and security of global maritime trade. Both have also proved willing to take an active, international role in countering the instability that followed the end of the Cold War's essentially bipolar world order. This is reflected in the acquisition of some powerful, flexible warships that provide valuable capabilities to the NATO alliance. However, financial considerations mean that their navies are now much smaller than they once were. It is by no means certain that there is enough money even to retain existing capabilities.

Whilst European events will always be a core interest to a British-based publication such as *World Naval Review*, it is important to note that it is tension in Asia that remains the most important influence on world naval developments. Much current focus is on the imminent outcome of the Philippines-initiated arbitration case against China's 'nine-dash line' claim in the South China Sea at the Permanent Court of Arbitration in The Hague, a hearing in which China has refused to participate. Both parties have been attempting to build diplomatic support for their position. A joint communique issued by the Association of South East Asian Nations (ASEAN) after a meeting with China's foreign minister Wang Yi on 14 June 2016 emphasising, *inter alia*, '… the importance of non-militarisation and self-restraint in the conduct of all activities, including land reclamation, which may have the potential to undermine peace, security and stability in the South China Sea …' was therefore something of a setback for China's diplomatic offensive given its current island-building strategy. Equally, the statement's swift retraction provides a strong indication

The changing global environment has meant that fleets that have reconfigured towards low-intensity stabilisation and peacekeeping activities may find themselves short of war-fighting assets. The Royal Netherlands Navy is one of a number of European fleets that find themselves in this position, albeit ships like the new JSS Joint Support Ship *Karel Doorman* pictured here appear sufficiently flexible to be useful in a range of scenarios. *(Royal Netherlands Navy)*

Table 1.0.2: MAJOR FLEET STRENGTHS MID 2016

REGION	**AMERICAS**			**EUROPE**						**ASIA**				**INDIAN OCEAN**	
Country	**Brazil**	**Canada**	**USA**	**France**	**Italy**	**Spain**	**Turkey**	**UK**	**Russia**	**Australia**	**China**	**Japan**	**Korea S**	**India**	**Pakistan**
Aircraft Carrier CVN/CV	1	–	10	1	1	–	–	–	1	–	1	–	–	1	–
Support Carrier CVS/CVH	–	–		–	1	–	–	–	–	–	–	3	–	–	–
Strategic Missile Sub SSBN	–	–	14	4	–	–	–	4	13	–	6	–	–	–	–
Attack Submarine SSGN/SSN	–	–	57	6	–	–	–	7	20	–	8	–	–	1	–
Patrol Submarine SSK	5	4	–	–	6	3	13	–	20	6	50	17	14	13	5
Battleships/Battlecruisers BB/BC	–	–	–	–	–	–	–	–	1	–	–	–	–	–	–
Fleet Escort CGN/CG/ DDG/FFG	8	13	85	17	18	11	16	19	25	11	60	37	24	24	10
Patrol Escort DD/FFG/FSG/FS	4	–	7	15	4	–	8	–	45	–	35	6	18	10	–
Missile Attack Craft PGG/PTG	–	–	–	–	–	–	20	–	35	–	75	6	17	10	8
Mine Countermeasures MCMV	5	12	11	14	10	6	15	15	40	6	25	27	9	6	3
Major Amp LHD/LPD/LPH/LSD	1	–	31	3	3	3	–	6	–	3	4	3	1	1	–

Notes:

1 Figures for Russia and China are approximate.

International relations in Asia continue to remain strained, partly because of China's assertive stance towards its maritime claims in the South China Sea and elsewhere. This is leading to growing co-operation between other nations that see their own interests threatened, albeit China's economic influence mean that some are unwilling to take an overtly confrontational stance. This image shows the Indian Project 17 frigate *Shakti* operating with other warships in the Philippine Sea in June 2016 during the latest Malabar training exercise with Japan and the United States. Japan was invited to become a permanent participant in the Malabar series in 2015. *(US Navy)*

of China's ability to put pressure on some ASEAN members to prevent emergence of a unified opposition.[8] This tension between fear of China's expansionary inclinations and acknowledgement of the realities of its economic influence is evident in Mrityunjoy Mazumdar's review of the Royal Malaysian Navy. As must surely be the case for other countries in the region, he concludes that Malaysia will face difficult choices in charting a way forward in the face of China's 'divide and rule' tactics.

SIGNIFICANT SHIPS

The naval significance of Asia is not, however, now limited to the impact of territorial disputes. The broader Indo-Pacific region is rapidly becoming a new hub of naval shipbuilding activity, a developing trend commented on in several previous editions of *World Naval Review*. The most recent country to benefit from this trend is India. Here, Garden Reach Shipbuilders & Engineers (GRSE) of Kolkata has emerged as preferred bidder for the Philippines new light frigate programme. If the contract – reportedly worth c. US$320m – is successfully concluded, GRSE will provide two ships based on the Project 28 *Kamorta* class anti-submarine design.[9]

One of the major drivers behind the *Kamorta* class programme was a desire to further the development of indigenous design and manufacturing capability, extending to localised supply of key components. This tendency is also apparent in Tomohiko Tada's review of Japan's *Akizuki* class destroyers which – whilst showing many US Navy influences – include significant amounts of locally-sourced equipment. Prominent amongst these is the Aegis-like Type 3A fire-control system and its associated FCS-3A multi-function radar, claimed to be amongst the most advanced of its type in the world. Another impressive Asian warship is the Republic of Korea Navy's *Dokdo* class amphibious assault ship, which is assessed by Guy Toremans. Now in commission for nearly ten years, she will shortly be joined by a sister-ship that is rumoured to have greater capabilities for fixed-wing aviation. The deployment of 'large-deck' aviation support ships – that also extends to Japan's *Hyuga* and *Izumo* helicopter-carrying destroyers, as well as the Australian *Canberra* class assault ships – is another trend amongst Asian fleets that reflects the growing capability of many of the leading regional navies.

The most significant new ship entering service during 2016 is undoubtedly the lead *Zumwalt* (DDG-1000) class destroyer. Conceived in the immediate aftermath of the Cold War, she was delivered on 20 May 2016 and her commissioning is scheduled for October. A futuristic, stealthy design with a pronounced tumblehome hull and ram-type bow that also gives her some resemblance to French pre-dreadnought battleships, she is subject to an extended review by Scott Truver and Edward Feege.[10] It is interesting to note that – in spite of her large size and considerable cost – she was initially intended to be a 'low end', littoral warfare focused warship complementing the multi-mission *Ticonderoga* (CG-47) and *Arleigh Burke* (DDG-51) classes. Whilst the curtailment of the *Zumwalt* class to just three ships means that they may prove to be something of a dead-end in US Navy warship construction, they incorporate a range of novel technologies that will undoubtedly exert a significant influence on future surface warships.

TECHNOLOGICAL DEVELOPMENTS

The balance between technological progress, available numbers and expense that is an underlying theme of a number of this year's chapters is also evident in David Hobbs' annual review of naval aviation. Notably, delays in fielding the advanced F-35 Lighting II Joint Strike Fighter is starting to create a 'fighter gap' that the US Navy can only resolve by continued construction of legacy types. The impact is evident in the US Department of the Navy FY2017 President's Budget Request, which – more broadly – continues the recent shift towards 'high intensity' war-fighting power at the expense of less sophisticated 'presence' assets, such as the littoral combat ship. Meanwhile, the British 2015 Strategic Defence and Security Review (SDSR) also confirms that future Royal Navy aviation will be increasingly focused on relatively limited numbers of new or modernised airframes.

The renewed danger that the US Navy may have to fight adversaries with advanced war-fighting potential – 'the most significant shift in the future security environment … is a return to an era of great power competition' according to a budget briefing by Deputy Secretary of Defense Robert Work – is also driving increased interest in anti-ship missiles. As is made clear in Norman Friedman's review of these systems, the US Navy has largely neglected anti-surface warfare capabilities since the end of the Cold War. Consequently, the rather dated Harpoon missile that forms the core of the US Navy's inventory of anti-ship missiles has been increasingly overtaken by the latest designs of Russian origin. A number of initiatives are underway to plug the capability gap but it seems that it will be a decade or so before a definitive solution is in place.

In contrast to the picture of a world order that is under significant pressure from great power rivalries found in much of this year's edition, we conclude with a study of a subject where international collaboration has produced significant successes. The complex world of Submarine Escape, Rescue, Abandonment & Survivability (SMERAS) is described with admirable clarity by another first-time contributor, Jan Ziolo. He makes the valuable point that the speed of response essential to the successful recovery of the crew of a downed submarine has been significantly assisted by global cooperation. It can only be hoped that this approach will survive current tensions.

The newly-delivered *Independence* (LCS-2) variant Littoral Combat Ship *Jackson* (LCS-6) has started to undergo a series of shock trials. The greater risk of the US Navy having to confront 'near peer' competitors such as China and Russia has resulted in greater emphasis being placed on 'high end' assets such as strike fighters and guided missiles. As such, current plans envisage procurement of second-tier units like *Jackson* being curtailed. *(US Navy)*

SUMMARY

The last year has seen little let-up in the friction that has steadily emerged from China's determination to enforce its extensive maritime claims and Russia's desire to regain the status – and, at least, some of the territory – it surrendered at the end of the Cold War. The predictable result has been an increase in global defence expenditure as these two countries' potential adversaries respond to the challenge and look to strengthen their own war-fighting potential. From a naval perspective, it is important to note that the consequence is not limited solely to the prospect of higher spending but may also ultimately change the composition of a number of fleets. More specifically, the post-Cold War emphasis in many navies on assets suitable for lengthy stabilisation and constabulary missions in littoral waters may well be subject to a rebalancing back towards ships more-suitable for front-line roles.

The changed environment will inevitably take time to have a practical effect. In the meantime, however, the fruits of programmes originated in the immediate aftermath of the Cold War are now coming to fruition. *Zumwalt* is, perhaps, the most spectacular result of these efforts to date but other new ships are following closely behind. In particular, the new carriers *Gerald R. Ford* (CVN-78) and *Queen Elizabeth* are due for delivery within the next twelve months. Their arrival – and other major developments – will undoubtedly be areas of focus for *World Naval Review 2018*.

An F/A-18E Super Hornet assigned to the 'Eagles' of Strike Fighter Squadron VFA-115 seen on the flight deck of the carrier *Ronald Reagan* (CVN-76) in June 2016. The US Navy has extended production of the Super Hornet to help alleviate a 'fighter gap' pending delivery of the next generation F-35C carrier variant of the Lightning II Joint Strike Fighter. *(US Navy)*

ACKNOWLEDGEMENTS

The continued publication of the *World Naval Review* series into its eighth year owes much to the support of a wide range of people, most notably a loyal readership and a growing list of contributors. It has been especially encouraging to welcome three new writers to this edition. Particular thanks must also go to Publishing Editor Robert Gardiner for his continued backing of the title, to Steve Dent for the high standards of his design work, and to John Jordan for the quality of his line drawings. I am also grateful to Hartmut Ehlers, Derek Fox, Bruno Huriet, Marc Piché, Haydn Sutton and Devrim Yayali for their supply of some most interesting photographs. In this regard, however, a special mention must go to Chris Cavas of *Defense News*, who generously provided a number of unique interior views of *Zumwalt* to give life to our extended coverage of the design. From industry, Marion Bonnet of DCNS, Esther Benito Lope of Navantia, Craig Taylor of Rolls-Royce and Matt Wickenheiser of General Dynamics/Bath Iron Works have all gone beyond the requirements of duty in their support. Last, but by no means least, the work of my wife Susan in undertaking initial proof reading in spite of an increasingly busy professional life is acknowledged with affection.

Comments and criticisms from readers are always appreciated; please direct them for my attention to info@seaforthpublishing.com

Conrad Waters, Editor, 30 June 2016

Notes

1. See Australia's *2016 Defence White Paper* (Canberra, Commonwealth of Australia, 2016) – currently downloadable at www.defence.gov.au/whitepaper/Docs/2016-Defence-White-Paper.pdf – and the British *National Security Strategy and Strategic Security and Defence Review 2015: Cm 9161* (London: Her Majesty's Stationery Office, 2015), which is currently available at: www.gov.uk/government/uploads/system/uploads/attachment_data/file/478933/52309_Cm_9161_NSS_SD_Review_web_only.pdf

2. This conclusion was reached, amongst others, by Benjamin David Baker in 'What Does Australia's Defense White Paper Actually Say?', *The Diplomat* – 5 March 2016 (Tokyo: The Diplomat, 2016).

3. See Sam Perlo-Freeman, Aude Fleurant, Pieter Wezeman and Siemon Wezeman, *SIPRI Fact Sheet April 2016: Trends In World Military Expenditure 2015* (Stockholm: SIPRI, 2016).

4. The impact of sanctions on Russia's naval renewal programme has been covered extensively, for example by Steven Beardsley in 'Sanctions put a crimp on Russia's naval plans', *Stars and Stripes* – 2 August 2015 (Washington DC: US Department of Defense, 2015).

5. See Steven Erlanger, 'Britain Votes to Leave E.U.; Cameron Plans to Step Down', *The New York Times* – 23 June 2016 (New York: The New York Times Company, 2016).

6. Although Project SEA 1000 dates from 2007, it was not until the defence white paper released in 2009 that Australia decided to double its submarine flotilla from six to twelve boats.

7. The factors influencing the wholesale reconfiguration undertaken by many navies after the end of the Cold War is somewhat beyond the scope of this annual. For a much more detailed assessment, please refer to the editor's *Navies in the 21st Century* (Barnsley: Seaforth Publications, 2016).

8. For further analysis see Tan Qiuyi and Saifulbari Ismail's 'Indonesia, Vietnam stress ASEAN's common stance on South China Sea', reported on *Channel NewsAsia's* website on 16 June 2016. Meanwhile a good overview of China's perspective on the Philippines initiated arbitration is provided by Xie Feng, China's Ambassador to Indonesia in 'The dangerous arbitration of Beijing-Manila dispute', *The Jakarta Post* – 9 June 2016 (Jakarta: PT Niskala Media Tenggara, 2016).

9. See Vivek Raghuvanshi's 'GRSE to Export Light Frigates to Philippines', *Defense News* – 12 May 2016 (Springfield VA: Sightline Media Group, 2016).

10. *Zumwalt's* first commanding officer is, perhaps appropriately, Captain James Kirk USN, namesake of the commanding officer of the fictional starship *Enterprise* (NCC-1701).

2.1 REGIONAL REVIEW

NORTH AND SOUTH AMERICA

Author:
Conrad Waters

INTRODUCTION

As always, naval developments in the Americas are ultimately dominated by financial and political considerations in the United States. The latest Bipartisan Budget Act (BBA), of November 2015, provided a further measure of relief from the caps on defence and other spending established by the 2011 Budget Control Act. By taking a two-year view, the legislation also ensured a welcome level of certainty over Department of Defense (DOD) spending for both FY2016 and FY2017. Total approved defence spending for FY2016 was a headline figure of US$580bn, a year-on-year increase of US$20bn and only a little under the US$585bn proposed under that year's Presidential Budget Request. The likely settlement for FY2017 is tighter, with the deal agreed under the BBA limiting the proposed FY2017 DOD budget to a little-changed US$583bn.[1] This is lower than previous DOD assumptions, resulting in cuts to planned procurement and the imposition of other efficiencies. The current administration has made it clear that further relaxation of financial controls will be needed after FY2017 if its defence strategy is to be achieved. Clearly, the outcome of the forthcoming Presidential Election will have a material impact on what actually transpires.

A key focus of the FY2017 budget was the perceived growth in the threat posed by China and Russia. They were explicitly named – alongside North Korea, Iran and global terrorism – as the five key challenges facing the US Military. As mentioned in the Introduction, this is resulting in priority being given to conventional deterrence against these two 'most advanced adversaries'.[2] For the US Navy, this has involved an increased focus on high-intensity war-fighting rather than lower-intensity presence capabilities, with a particularly adverse impact on future Littoral Combat Ship (LCS) numbers. Planned acquisitions of the type will fall from fifty-two to forty, protecting investment in areas such as naval aircraft and munitions that are perceived as being more critical in the current environment.

The new plan halves LCS purchases over the next five years and starts to take effect immediately. The change is one of the factors behind a c. ten per cent fall in the previously envisaged FY2017 shipbuilding budget to US$14.7bn. Overall headline US Navy spending is budgeted at US$165bn, around US$4bn less than that enacted for the previous year.[3] This will also have a modest impact on the number of sailors. This will stabilise slightly below previously-planned levels at c. 323,000 active personnel. In spite of the decision to reduce LCS numbers, forward presence remains a key objective. Indeed, numbers of deployed ships are expected to increase over the next five years as completion of ships ordered in previous years reverses the recent decline in fleet size.

Meanwhile, to the north, Canada's new Liberal government has embarked on what has been described as that country's first comprehensive defence review since its 1994 Defence White Paper. The review is due to be concluded early in 2017. Many commentators believe that its findings will favour naval equipment procurement over other projects, particularly the Royal Canadian Air Force's potential acquisition of F-35A Lightning II Joint Strike Fighters. It is likely that the recent focus on strengthening the military's presence in the Canadian Arctic – primarily a naval/coast guard task – will also be maintained. However, it seems apparent that Canada is unlikely to follow the trend of rising defence spending seen elsewhere in the world, with the budget likely to be fixed at the current level of c. one per cent of GDP.

As such, there must be considerable uncertainty whether the navy's future aspirations – most significantly its plan for up to fifteen new Canadian Surface Combatants to replace existing *Iroquois* and *Halifax* class vessels – will be realised in their entirety. Certainly, recent reports that the likely construction cost of the new ships could more than double to CAD$30bn (US$23bn) is not encouraging.[4] The new administration remains committed to the National Shipbuilding Strategy that was first announced in 2010 but is already taking steps to 'streamline' the surface combatant programme. On 13 June 2016 it was announced that Canada would modify an existing warship design rather than continue with the previous approach of selecting a warship designer and a combat systems integrator to produce a bespoke design. Whether this will be sufficient to provide the required numbers remains a matter of conjecture.

Canada is currently in the middle of its first comprehensive defence review since 1994. The navy hopes the outcome will support its procurement objectives, notably the Canadian Surface Combatant that will replace the *Iroquois* and *Halifax* classes. This image shows the modernised *Halifax* class frigate *Winnipeg* exercising with the Chilean replenishment tanker *Almirante Montt* in May 2016; past procurement delays mean that the Royal Canadian Navy does not currently have any replenishment vessels of its own. *(Canadian Armed Forces)*

Table 2.1.1: FLEET STRENGTHS IN THE AMERICAS – LARGER NAVIES (MID 2016)

COUNTRY	ARGENTINA	BRAZIL	CANADA	CHILE	COLOMBIA	ECUADOR	PERU	USA
Aircraft Carrier (CVN/CV)	–	1	–	–	–	–	–	10
Strategic Missile Submarine (SSBN)	–	–	–	–	–	–	–	14
Attack Submarine (SSN/SSGN)	–	–	–	–	–	–	–	57
Patrol Submarine (SSK)	3	5	4	4	4	2	6	–
Fleet Escort (CG/DDG/FFG)	4	8	13	8	4	2	8	85
Patrol Escort/Corvette (FFG/FSG/FS)	9	4	–	–	1	6	–	7
Missile Armed Attack Craft (PGG/PTG)	2	–	–	3	–	3	6	–
Mine Countermeasures Vessel (MCMV)	–	5	12	–	–	–	–	11
Major Amphibious Units (LHD/LPD/LPH/LSD)	–	1	–	1	–	–	–	31

MAJOR NORTH AMERICAN NAVIES – CANADA

The future structure of Canada's armed forces depends on the outcome of the current defence review. However, elements of the 2010 National Shipbuilding Strategy are already making progress. Notably, construction is now underway on the new, ice-capable Arctic offshore patrol ships of the *Harry DeWolf* class. Initial steel cutting took place at Irving Shipbuilding's Halifax Shipyard in June 2015 followed by the start of full construction on 1 September of that year. The new ships are intended to strengthen the enforcement of Canadian sovereignty in Arctic waters and are scheduled for delivery from 2018 onwards. Displacing around 6,500 tons – considerably more than the *Halifax* class frigates – the 103m-long ships are lightly armed with a BAE Systems 25mm Mk 38 gun to support constabulary operations. Of greater significance is their ability to carry a wide range of payloads, including shipping containers, underwater survey equipment or even a landing craft. There is also a special vehicle bay to house trucks, all-terrain vehicles and snowmobiles. A 20-ton capacity crane supports loading/unloading operations, whilst there are facilities to accommodate an embarked helicopter of up to CH-148 Cyclone size. A diesel-electric propulsion system provides a maximum speed in the order of 17 knots. It is hoped that the CAD$3.5bn (US$3.5bn) budget allocated to the programme will allow the completion of six ships. However, some reports suggest funding might not stretch far enough to pay for the as-yet unnamed sixth vessel. Work has also started on new refuelling and support facilities at Nanisivik on Baffin Island to provide logistic facilities for the class.

Meanwhile, preparations are underway for commencement of work on the first of two *Queenston* class joint support ships. These are to be built by Seaspan's Vancouver Shipyards on the west coast to a design based on the German Type 702 *Berlin* class. Considerable delays in finalising the estimated CAD$2.6bn (US$2bn) programme – it was originally announced as long ago as 2004 – have meant that the Royal Canadian Navy's existing *Protecteur* class replenishment oilers have already been retired. Agreements with Chile and Spain for temporary access to their own vessels has eased some of the strain. However, a more comprehensive solution has been found under Project Resolve. This will involve a consortium headed by Quebec's Chantier Davie yard converting the German-built container ship *Asterix* to act as an interim auxiliary oiler replenishment ship from 2017 until the second of the *Queenston* class vessels is delivered in 2021. The agreement involves the consortium funding both the capital cost of acquisition/conversion and ongoing crewing/operating charges in return for payments of c. CAD$700m (US$550m) over the life of a planned five-year deal.[5]

Although the use of a modified foreign design is expected to speed delivery of the Canadian Surface Combatant programme by two years, it will not be

Table 2.1.2: CANADIAN NAVY: PRINCIPAL UNITS AS AT MID 2016

TYPE	CLASS	NUMBER	TONNAGE	DIMENSIONS	PROPULSION	CREW	DATE
Principal Surface Escorts							
Destroyer – DDG	IROQUOIS	1	5,100 tons	130m x 15m x 5m	COGOG, 29 knots	280	1972
Frigate – FFG	HALIFAX	12	4,800 tons	134m x 16m x 5m	CODOG, 29 knots	225	1992
Submarines							
Submarine – SSK	VICTORIA (UPHOLDER)	4	2,500 tons	70m x 8m x 6m	Diesel-electric, 20+ knots	50	1990

Two computer-generated images of the new *Harry DeWolf* class Arctic offshore patrol ships on which construction has now started at Irving Shipbuilding's Halifax Shipyard. The lightly-armed, ice-strengthened vessels are intended to perform presence and constabulary duties in Canada's Arctic, a region of increased strategic importance. *(Irving Shipbuilding Inc.)*

The decommissioned replenishment oiler *Protecteur* pictured departing the naval base at Esquimalt on 24 February 2016 on her final journey to the scrapyard. She suffered a major engine room fire in February 2014 and it was determined repair was uneconomic given her limited remaining service life. Canadian west-coast shipyards declined to bid for the scrapping contract and the ship had to be towed 7,600 miles to a yard in Nova Scotia for the work to be carried out. *(Canadian Armed Forces)*

until construction of the *Harry DeWolf* class begins to wind down at Halifax that assembly of the new ships will commence.[6] In the meantime, surface fleet operations will remain in the hands of the existing *Halifax* class, which have been modernised under the *Halifax* Class Modernisation/Frigate Equipment Life Extension (HCM/FELEX) programme. In April 2016, an important project milestone was achieved with completion of work on *Regina*, the fifth and final frigate to undergo modernisation at Seaspan's west-coast Victoria Shipyards. The upgrade of *Toronto*, the seventh and last vessel to be modernised at the east-coast Halifax Shipyard is expected to be finished during the summer of 2016, with the programme on schedule to be finalised in accordance with its expected timescale at the start of 2018. A number of the upgraded ships have already deployed operationally and on exercises. Most significantly, both *Winnipeg* and *Fredericton* have operated with Standing NATO Maritime Group Two as part of Operation 'Reassurance', the Canadian military's support for NATO assurance measures following Russia's intervention in Ukraine.

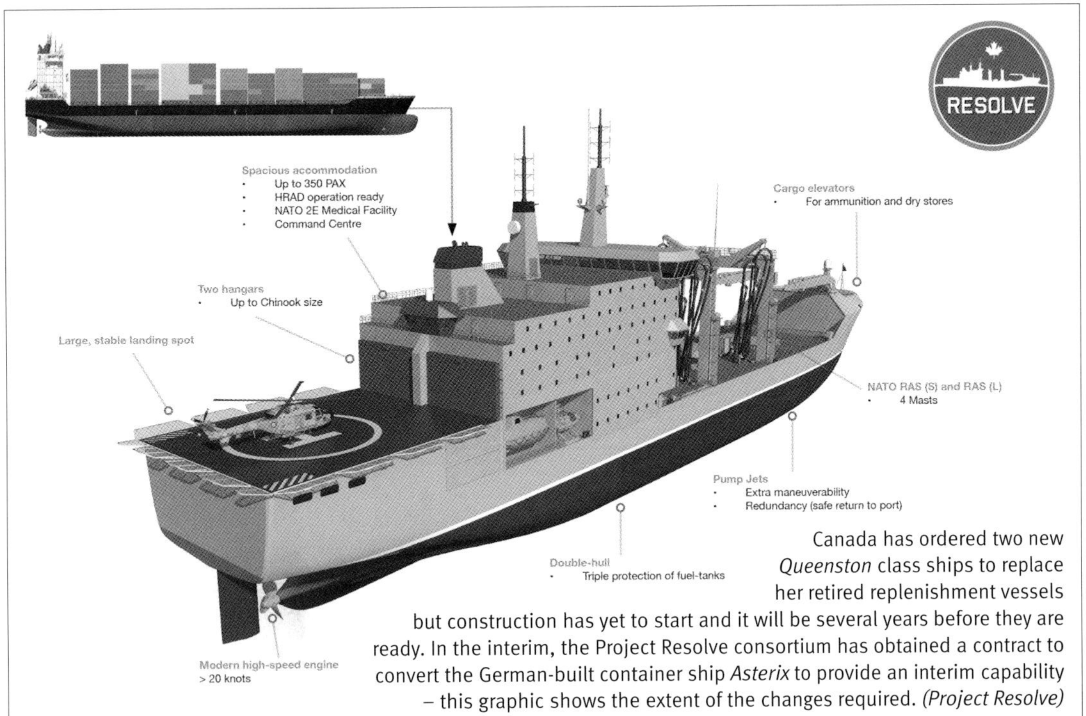

Canada has ordered two new *Queenston* class ships to replace her retired replenishment vessels but construction has yet to start and it will be several years before they are ready. In the interim, the Project Resolve consortium has obtained a contract to convert the German-built container ship *Asterix* to provide an interim capability – this graphic shows the extent of the changes required. *(Project Resolve)*

The December 2014 announcement of the achievement of fully operational status for the four-strong *Victoria* (former British *Upholder*) class submarine flotilla has proved somewhat premature. The discovery of defective welds undertaken by a sub-contractor to the *Victoria* in-Service Support Contract (VISC) mean that the west-coast based submarines *Chicoutimi* and *Victoria* herself will need significant rectification work lasting into 2017. With *Corner Brook* in planned maintenance, this leaves only the east-coast based *Windsor* fully available for tasking at the present time.

MAJOR NORTH AMERICAN NAVIES – UNITED STATES

As demonstrated by Table 2.1.3, US Navy Battle Force size had stabilised at c.270–275 ships as of June 2016. The expectation is that fleet numbers will experience a sustained rise over the rest of the decade. This is due largely to previous orders for significant numbers of Littoral Combat Ships (LCS) and, also, the restart of *Arleigh Burke* (DDG-51) class destroyer production. The US Navy currently maintains a force structure requirement of 308 ships. This is likely to be achieved in total – but not in terms of all constituent ship categories – during

FY2021. The longer-term prognosis is less positive, with plans indicating shortfalls of surface combatants and attack submarines during the 2030s due to curtailment of LCS/light frigate construction and the need to complete the *Ohio* Replacement class (formerly SSBN-X) strategic submarines.[7] It is also noteworthy that even this likely battle force inventory requires funding that the US Navy has historically committed to new warship construction. A further complication is that the 308-ship requirement is based on a force structure assessment last carried out in 2012 and that was subject to only an interim update in 2014. The assumptions behind the requirement are now being revisited in light of the changed strategic environment. Press reports suggest that both major surface combatant and attack submarine goals are likely to be revised upwards.[8] Conversely, it seems that targeted numbers of minor surface combatants will move downwards given the lower LCS requirement.

A major impact on future fleet size will inevitably be the size of the future shipbuilding programme. Although this has remained inadequate to support longer-term goals, the navy has been reasonably successful in increasing the numbers of ships acquired in recent years. The headline numbers in the latest FY2017–FY2022 procurement plan – highlighted in Table 2.1.4 – are therefore somewhat disappointing. They show a ten-ship reduction (from forty-eight to thirty-eight vessels) when compared with the equivalent FY2016–FY2021 programme. Whilst part of this is a reflection of Congressional acceleration of an expeditionary mobile base and a fleet tug into FY2016, the main factor is a halving of proposed LCS/light frigate orders from fourteen to seven units. Also noteworthy is the halving of orders for *Virginia* (SSN-774) class submarines to just one boat in FY2021 as the *Ohio* Replacement Submarine programme begins in earnest.

The current composition of the US Navy's principal battle force units is set out in Table 2.1.5, whilst further details on major ship categories follow.

Table 2.1.3: US NAVY BATTLE-FORCE DEVELOPMENT (FY 2015 TO FY 2021)

SHIP TYPE	GOAL	MID-2015	MID-2016	SEP 2017	SEP 2018	SEP 2019	SEP 2020	SEP 2021
Aircraft Carrier (CVN)	11	10	10	11	11	11	11	11
Strategic Submarine (SSBN)	12	14	14	14	14	14	14	14
Attack Submarine (SSGN/SSN)	48	57	57	56	57	56	56	55
Fleet Escort (CG/DDG)	88	84	85	90	91	94	95	97
Patrol Escort (FF/LCS/MCMV)	52	18	18	25	29	32	33	34
Amphibious Vessel (LHA/LHD/LPD/LSD)	34	30	31	32	32	33	33	33
Other – Logistics & Support Ships	63	59	60	59	61	60	64	64
TOTAL	308	272	274	287	295	300	306	308

Note

Forward data is derived from the *Report to Congress on the Annual Long-Range Plan for Construction of Naval Vessels – June 2014* and *Report to Congress on the Annual Long-Range Plan for Construction of Naval Vessels – March 2015*, including their appendices on planned ship decommissionings. The numbers shown for 30 Sep 2015 reflect delays to delivering *Zumwalt* (DDG-1000); in addition, it seems unlikely a further four LCS will have been commissioned by then.

Table 2.1.4: USN FY2017 FIVE YEAR SHIPBUILDING PLAN (FY2017–FY2021)

SHIP TYPE	FY2016: ACTUAL	FY2017: REQUEST	FY2018: PLAN	FY2019: PLAN	FY2020: PLAN	FY2021: PLAN
Aircraft Carrier (CVN-78)	Nil (Nil)	Nil (Nil)	1 (1)	Nil (Nil)	Nil (Nil)	Nil (Nil)
Strategic Submarine (SSBN-X)	Nil (Nil)	Nil (Nil)	Nil (Nil)	Nil (Nil)	Nil (Nil)	1 (1)
Attack Submarine (SSN-774)	2 (2)	2 (2)	2 (2)	2 (2)	2 (2)	1 (1)
Destroyer (DDG-51)	2 (2)	2 (2)	2 (2)	2 (2)	2 (2)	2 (2)
Littoral Combat Ship (LCS-1/2)	3 (3)	2 (3)	1 (3)	1 (2)	1 (3)	2 (3)
Amphibious Assault Ship (LHA-6)	Nil (Nil)	1 (1)	Nil (Nil)	Nil (Nil)	Nil (Nil)	Nil (Nil)
Amphibious Transport Ship (LPD-17)	1 (1)	Nil (Nil)	Nil (Nil)	Nil (Nil)	Nil (Nil)	Nil (Nil)
Amphibious Ship (LX-R)	Nil (Nil)	Nil (Nil)	Nil (Nil)	Nil (Nil)	1 (1)	Nil (Nil)
Joint High Speed Vessel (JHSV-1)	1 (Nil)	Nil (Nil)	Nil (Nil)	Nil (Nil)	Nil (Nil)	Nil (Nil)
Mobile Landing Platform (MLP-1)	1 (Nil)	Nil (1)	Nil (Nil)	Nil (Nil)	Nil (Nil)	Nil (Nil)
Replenishment Oiler (TAO(X))	1 (1)	Nil (Nil)	1 (1)	1 (1)	1 (1)	1 (1)
Fleet Tug (TATS)	1 (Nil)	Nil (1)	1 (1)	1 (2)	1 (1)	1 (Nil)
Total	12 (9)	7 (10)	8 (10)	7 (9)	8 (10)	8 (8)

Notes:

1 Figures in brackets relate to previous FY2016 Budget Request and Shipbuilding Plan.

2 Figures relate to Ship Battle Forces – Other Ships are not included.

3 LCS production is planned to transition to the upgraded FF in FY2019.

4 The FY2016 MLP mobile landing platform ship will be built as an afloat forward staging base (AFSB) variant.

5 The SSBN-X is now known as the *Ohio* Replacement Submarine; the JHSV is now known as the Expeditionary Fast Transport (EPF) and the MLP AFSB is now the Expeditionary Mobile Base (ESB).

6 The FY2016 shipbuilding programme increased due to Congressional addition of one JHSV, one MLP and one TATS.

Aircraft Carriers: The lead *Gerald R. Ford* (CVN-78) class carrier is currently expected to commence sea trials in July 2016 following further programme delays that appear to be related to the innovative advanced arresting gear (AAG) that is being fitted to the ship. Although delivery is planned for the end of September 2016, it appears that the AAG will not be cleared to support fixed-wing operations until later in the year, when sea-based trials with F/A-18E/F Super Hornets will commence. There have been reports that the US Navy might revert to an upgraded variant of its current Mk 7 Mod 3 system for the two follow-on ships in the class as a result of the cost of AAG more than doubling. This is not, however, a straightforward decision as AAG is better suited to handling a

Table 2.1.5: UNITED STATES NAVY: PRINCIPAL UNITS AS AT MID 2016

TYPE	CLASS	NUMBER	TONNAGE	DIMENSIONS	PROPULSION	CREW	DATE
Aircraft Carriers							
Aircraft Carrier – CVN	**NIMITZ** (CVN-68)	10	101,000 tons	340m x 41/78m x 12m	Nuclear, 30+ knots	5,700	1975
Principal Surface Escorts							
Cruiser – CG	**TICONDEROGA** (CG-47)	22	9,900 tons	173m x 17m x 7m	COGAG, 30+ knots	365	1983
Destroyer – DDG	**ZUMWALT** (DDG-1000)	1	15,800 tons	186m x 25m x 8m	IEP, 30+ knots	175	2016
Destroyer – DDG	**ARLEIGH BURKE** (DDG-51) – Flight II-A	34	9,400 tons	155m x 20m x 7m	COGAG, 30 knots	380	2000
Destroyer – DDG	**ARLEIGH BURKE** (DDG-51) – Flights I/II	28	8,900 tons	154m x 20m x 7m	COGAG, 30+ knots	340	1991
Littoral Combat Ship – FS	**FREEDOM** (LCS-1)	3	3,500 tons	115m x 17m x 4m	CODAG, 45+ knots	<50[1]	2008
Littoral Combat Ship – FS	**INDEPENDENCE** (LCS-2)	4	3,000 tons	127m x 32m x 5m	CODAG, 45+ knots	<50[1]	2010
Submarines							
Submarine – SSBN	**OHIO** (SSBN-726)	14	18,800 tons	171m x 13m x 12m	Nuclear, 20+ knots	155	1981
Submarine – SSGN	**OHIO** (SSGN-726)	4	18,800 tons	171m x 13m x 12m	Nuclear, 20+ knots	160	1981
Submarine – SSN	**VIRGINIA** (SSN-774)	12	8,000 tons	115m x 10m x 9m	Nuclear, 25+ knots	135	2004
Submarine – SSN	**SEAWOLF** (SSN-21)	3[2]	9,000 tons	108m x 12m x 11m	Nuclear, 25+ knots	140	1997
Submarine – SSN	**LOS ANGELES** (SSN-688)	38	7,000 tons	110m x 10m x 9m	Nuclear, 25+ knots	145	1976
Major Amphibious Units							
Amph. Assault Ship – LHD	**AMERICA** (LHA-6)	1	45,000 tons	257m x 32/42m x 9m	COGAG, 20+ knots	1,050	2014
Amph Assault Ship – LHD	**WASP** (LHD-1)	8[3]	41,000 tons	253m x 32/42m x 9m	Steam, 20+ knots	1,100	1989
Landing Platform Dock – LPD	**SAN ANTONIO** (LPD-17)	10	25,000 tons	209m x 32m x 7m	Diesel, 22+ knots	360	2005
Landing Ship Dock – LSD	**WHIDBEY ISLAND** (LSD-41)	12[4]	16,000 tons	186m x 26m x 6m	Diesel, 20 knots	420	1985

Notes:

1 Plus mission-related crew. **2** Third of class, SSN-23 is longer and heavier. **3** LHD-8 has many differences. **4** Includes four LSD-49 HARPERS FERRY variants.

broader range of aircraft, including the next-generation lighter unmanned aerial vehicles.[9]

Other new systems installed in CVN-78, including its EMALS electro-magnetic catapults and dual band radar, appear to be faring well. EMALS, particularly, is proving more reliable than predicted. The new carrier is expected to embark on a protracted series of trials, training and work-up after delivery, culminating in shock trials in 2019. As such, it is unlikely to be before the end of the decade that she commences operational deployment. Meanwhile, work on the second member of the class – *John F. Kennedy* (CVN-79) – is around twenty per cent complete following a formal keel-laying ceremony in August 2015. The first advanced procurement funding for the next ship, *Enterprise* (CVN-80) was authorised in the FY2016 budget prior to a request for formal approval in FY2018.

Surface Combatants: The near-four year hiatus in deliveries of major surface combatants ended in May 2016 with the transfer of the first *Zumwalt* (DDG-1000) class destroyer to navy custody at the Bath Iron Works (BIW) yard in Maine. Her sister,

The new US Navy aircraft carrier *Gerald R. Ford* (CVN-78) is now close to completion and is scheduled to commence sea trials in July 2016. This picture shows her being turned in the James River, Virginia on 11 June 2016 as part of pre-trial preparations. *(US Navy)*

The new Flight IIA *Arleigh Burke* class destroyer *Ralph Johnson* (DDG-114) seen being floated out by HII's Ingalls yard in December 2015. Ships from the restarted *Burke* class programme will start to be delivered over the next twelve months. *(Huntington Ingalls Industries)*

An unusual view of the new *Independence* variant LCS *Omaha* (LCS-12) being transferred on a barge down the Mobile River to a floating-out dock on 19 November 2015. The water jets that form a key part of her propulsion system are clearly visible. Overall numbers of US Navy ships are set to rise over the next few years, with the delivery of a significant number of LCS a key factor behind this growth. *(Austal USA)*

Michael Monsoor (DDG-1001), was christened on 18 June 2016. She is expected to be delivered before the end of 2017, although this may be subject to delay. There were a number of reports towards the end of 2015 that consideration was being given to cancelling the third ship, *Lyndon B. Johnson* (DDG-1002), on cost grounds.[10] However, perhaps because construction is well-advanced, the proposal now appears to be off the agenda. Further detail on the broader *Zumwalt* class programme is contained in Chapter 3.3.

With the *Zumwalt* class truncated to just three ships, future US Navy surface-combatant construction is dominated by renewed fabrication of *Arleigh Burke* class destroyer variants. All of the additional twelve Flight IIA type ships envisaged under the restarted programme have now been ordered.[11] Of these, *John Finn* (DDG-113) is close to completion at Huntington Ingalls Industries' (HII's) Ingalls yard, which also launched *Ralph Johnson* (DDG-114) on 12 December 2015. Construction is being shared with Bath Iron Works, which floated out *Rafael Peralta* (DDG-115) the previous November. BIW is also being given responsibility for the first, modified Flight III variant. This incorporates Raytheon's new modular AN/SPY-6 air and missile-defence radar (AMDR), as well as increased electrical generation capacity and an upgraded electrical grid. Further Flight III vessels will follow at a rate of two p.a. over the FY2017–FY2022 planning period.

The other main constituent of the surface fleet are the remaining twenty-two *Ticonderoga* (CG-47) class cruisers, which first introduced the Aegis system into operational service in 1982. The class have better command and control facilities than the smaller destroyers and one is typically assigned to each carrier group to co-ordinate the group's air defence. The oldest eleven of these cruisers have already been upgraded to the latest Aegis standard. The US Navy now wants to undertake a phased modernisation of the remaining ships, using them to replace the older ships on a one-for-one basis as they fall due for retirement from FY2020 onwards. This would involve putting some of the unmodernised ships into temporary reserve to preserve their hull lives until upgrade commenced. There has been significant Congressional resistance to the plan, largely due to concerns that the ships put into reserve would never be reactivated. A compromise deal has limited the number of ships that can enter the modernisation programme at any one time.

However, the navy is attempting to revisit the issue in the FY2017 budget. It argues that its original proposal would be more cost-effective and better ensure maintenance of an adequate fleet of command ships until a new generation of surface combatants enters service in the 2030s.

Turning to smaller surface combatants, the decommissioning of *Simpson* (FFG-56) at Mayport in Florida on 29 September 2015 marked the end of US Navy service for the *Oliver Hazard Perry* (FFG-7) class frigates that first joined the fleet in 1977. Deliveries of both variants of the new LCS that serve as their effective replacements are starting to ramp up. *Milwaukee* (LCS-5) and *Jackson* (LCS-6) both commissioned towards the end of 2015 and *Montgomery* (LCS-8) was delivered on 23 June 2016. Orders for the type now extend to thirteen of each variant. Under a previous programme restructuring announced in 2014, production was to shift to an upgraded, multi-mission light frigate configuration with effect from LCS-33 in FY2019. This would have given a total of thirty-two of the basic design and twenty of the frigates. However, the subsequent December 2015 decision to truncate the programme and slow construction means that the switch to the revised design will now seemingly take effect from the twenty-ninth ship. This produces a revised total of twenty-eight LCS and twelve frigates. The intention is also to focus on either the *Freedom* (LCS-1) or *Independence* (LCS-2) hull form at that stage, with inevitable consequences for which one of current builders Marinette Marine and Austal USA loses out. Two *Freedom* class ships – *Fort Worth* (LCS-3) and the new *Milwaukee* – have suffered high-profile engineering breakdowns in the past twelve months; these will not have helped that variant's cause.

The new frigates will have greater basic war-fighting potential than the earlier ships but less capacity to accept the specialised mission modules that are a key part of the initial LCS concept. Moreover, it also seems that previous emphasis on the rapid interchangeability of these modules during deployments may be downgraded in favour of a 'one ship, one mission' approach for ships built to the basic design. Under this approach, these ships would be deployed with a specific anti-submarine, surface warfare or mine-countermeasures modules and be crewed accordingly.[12]

Amphibious Shipping: The US Navy's front-line amphibious force has been increased to thirty-one ships with the delivery of the tenth *San Antonio* (LPD-17) amphibious transport dock, *John P. Murtha* (LPD-26), on 13 May 2016. She will be formally commissioned in the autumn. The eleventh unit of the class, *Portland* (LPD-27), was launched from HII's Ingalls yard on 13 February 2016. The programme will conclude with a twelfth and final unit, to be named *Fort Lauderdale* (LPD-28). The ship was formally approved in the FY2016 programme after being allocated unrequested funding by Congress in previous years.[13] She will omit the composite masts of previous members of the class and is being seen as a bridge to the new LX(R) design that will be ordered from FY2020 onwards. Industry would like to bring this date forward to achieve the benefits of continuity of production; this will ultimately depend on the budgetary position and, doubtless, the continued strength of HII's lobbying efforts. In the meantime, a third – as yet unnamed – member of the *America* (LHA-6) class amphibious assault ships is included in the FY2017 budget request. Her design includes the addition of a well deck omitted from the first two ships.

A view of the *Freedom* variant LCS *Milwaukee* (LCS-5) transiting the St Lawrence Seaway on 30 November 2015, shortly after her commissioning at Milwaukee on 21 November. The ship subsequently suffered complete propulsion failure on 11 December after damage to a clutch and had to be towed to port. The damage has now been repaired. *(Marc Piché)*

The long-term amphibious fleet goal of thirty-four ships still remains below that required to meet the stated objective of being able to transport the assault echelons of two Marine Expeditionary Brigades, let alone additional requirements that may arise for amphibious shipping. One way of squaring this particular circle has been investment in low-cost derivatives of civilian vessels to act as auxiliary amphibious ships. These include the *Spearhead* (EFP-1) class Expeditionary Fast Transports (formerly designated Joint High Speed Vessels), *Montford Point* (ESD-1) class Expeditionary Transport Docks (the former Mobile Landing Platforms) and *Lewis B. Puller* (ESB-3) class Expeditionary Mobile Bases (previously designated Afloat Forward Staging Bases).[14] The ships have been well-received in service and orders have been increased in consequence. However, they inevitably lack the survivability of true front-line warships.

The successful conduct of amphibious operations clearly requires effective transport from ship to shore over distances that are increasing due to the greater availability of precision weapons that can threaten major amphibious units. This is pushing an increased emphasis on air transportation – notably by the MV-22B Osprey and the CH-53 series of heavy lift helicopters – but surface connectors are still required. The most important current vessel in this category is the LCAC hovercraft, which are now approaching life expiry. The first examples of the replacement SSC ship-to-shore connectors are expected to be delivered for trials by Textron during 2017 before becoming operational in 2020. Preliminary design work is also underway on a surface connector replacement – SC(X)R – for the

The US Navy's amphibious fleet increased to thirty-one ships, still somewhat short of the desired level, with the delivery of the tenth LPD-17 type amphibious transport dock, *John P. Murtha* (LPD-26), in May 2016. This picture shows her on trials in March of that year. *(Huntington Ingalls Industries)*

The US Navy has been compensating for a shortfall in front-line amphibious vessels by innovative adaptations of commercial designs for second-line roles. This image shows *Lewis B. Puller* (ESB-3), the first of three expeditionary mobile bases that are designed to support Special Forces, mine-countermeasures operations and various other command and intelligence-gathering functions. She is pictured off Norfolk, Virginia in October 2015. *(Marc Piché)*

existing LCU 1600-class landing craft. Both types will be very similar in configuration to the types they replace but exploration of new concepts is also underway. The final piece of the jigsaw is the new AVC amphibious combat vehicle – a replacement for the venerable assault amphibious vehicles first introduced in 1972. BAE Systems and Science Applications International Corporation have been shortlisted to demonstrate competing off-the-shelf designs prior to the award of a contract for an interim vehicle. A more ambitious, bespoke design will be fielded at a later date.

Submarines: Construction of Block IV *Virginia* class submarines continues at the rate of two p.a. under the latest multi-year procurement contract agreed for FY2014–FY2018. Production will then transition to the improved Block V variant. Most of these will carry the new Virginia Payload Module (VPM) that contains four additional large-diameter launch tubes for twenty-eight additional Tomahawk cruise missiles or similar weapons. This increase in weapons-carrying capacity will go some way to compensate for the withdrawal of the current converted *Ohio* (SSGN-726) cruise missile submarines, which can each carry up to 154 missiles. However, as previously referenced, the need to commence construction of *Ohio* Replacement class strategic submarines in the early 2020s will result in a fall in *Virginia* class procurement. Given the rate of retirement of the remaining Cold War *Los Angeles* (SSN-668) class boats, this will lead to a growing shortfall against the required number of attack submarines from the second half of the decade onwards. The US Navy is looking at ways to reduce the gap. Potential solutions include modest increases to the service lives of existing submarines; deployments of longer duration; and the possibility of finding sufficient efficiencies to afford an additional, second *Virginia* class boat in FY2021.[15] A further complication is the discovery of defective welding in three of the class, removing them from the operational fleet until complicated repairs can be completed.[16]

The *Ohio* Replacement programme, meanwhile, continues to move ahead. The navy announced in March 2016 that General Dynamics Electric Boat had, as expected, been selected as prime contractor for a project that could eventually amount to as much as US$100bn. In contrast to the arrangement for the *Virginia* class, where construction is split between Electric Boat and HII's Newport News facility, Electric Boat will undertake final assembly of all the strategic submarines. However, Newport News will be sub-contractor for around a fifth of the work and be given a larger share of future *Virginia* class orders.[17]

Turning to operations, the current emphasis on maintaining forward presence in key regional hotspots is inevitably having an impact on a fleet that is still some five years away from achieving its targeted size. Consequences of this have included greater length and unpredictability of deployments, a greater maintenance burden and an associated increased risk of gaps in 'coverage' emerging. A revised Optimized Fleet Response Plan (OFRP) was introduced in 2014 to try to bring greater stability to carrier strike group deployments for the benefit both of operational commands and the ships' crews. Under this 'supply-based' model, strike groups will carry out a single seven-month deployment in a fixed 36-month cycle. This will allow sufficient time for maintenance and training, as well as including an extended 'sustainment' phase based at the group's homeport to provide surge capacity. The jury remains out on the success of the new structure, although increases to maintenance budgets will certainly facilitate the fixed cycles working to schedule.

Another possible consequence of forward presence is a greater risk of clashes with potential adversaries. This danger was vividly demonstrated by a high profile incident in April 2016 where Russian Su-24 'Fencer' attack aircraft made repeated low level passes over the destroyer *Donald Cook* (DDG-75) whilst the US Navy was conducting exercises with Polish forces in the Baltic. There have been similar incidents involving US Navy reconnaissance aircraft and Chinese jets over the South China Sea, a region where American surface naval forces have also been stepping up their presence. A major deployment to the Philippine Sea in June 2016 saw two aircraft carriers – *John C. Stennis* (CVN-74) and *Ronald Reagan* (CVN-76) – operating together in a training exercise that drew strident criticism from the Chinese press. Potentially the most dangerous flashpoint, however, remains the Persian Gulf. Here two US Navy CB-90 riverine command boats and their ten-strong crew were briefly detained by Iran's Islamic Revolutionary Guard after a navigational error caused them to stray into Iranian waters. The recent agreement of an accord on Iran's nuclear

The *Los Angeles* class submarine *Springfield* (SSN-761) pictured on a port visit to the Royal Navy submarine base at Faslane, near Glasgow in April 2016. The Cold War-era *Los Angeles* class boats are set to be decommissioned faster than replacement *Virginia* class submarines can be built, producing a gap in US Navy submarine capabilities. *(US Navy)*

The US Navy *Nimitz* class aircraft carriers *John C. Stennis* (CVN-74) and *Ronald Reagan* (CVN-76) pictured operating together in the Philippine Sea in June 2016 as part of a training exercise that was criticised in the Chinese press. The Optimized Fleet Response Plan (OFRP) introduced in 2014 is set to give greater predictability to US Navy aircraft carrier availability. *(US Navy)*

programme probably spurred a swift resolution to a situation that could have had serious consequences.

The US Coast Guard continues to take delivery of 'Legend' class national security cutters and smaller 'Sentinel' class fast response cutters in replacement for legacy *Hamilton* (WHEC-715) and 'Island' class types. Five 'Legend' class cutters are now in service following the commissioning of *James* (WMSL-754) in August 2015. Congress funded an unplanned ninth member of the class in FY2016 and there is also some Congressional pressure for a tenth ship. Thirty-eight fast response cutters have been approved to date out of a planned total of fifty-eight. All will be built by Bollinger Shipyards of Louisiana following a May 2016 following conclusion of a tender process for the final twenty-six. The result of competitive bids for the design of an intermediate offshore patrol cutter class – which is eventually planned to total twenty-five vessels – is expected to be announced before the end of 2016.

OTHER NORTH AND CENTRAL AMERICAN NAVIES

To the south of the United States, **Mexico** continues to prioritise constabulary missions for its naval forces, which have played a significant and largely effective role in the fight against organised crime. The Mexican Naval Infantry Force has both sustained and inflicted casualties in actions against the powerful drug cartels, playing a major role in the recapture of the notorious drugs baron Joaquin 'El Chapo' Guzman at the start of 2016. The emphasis on policing and maritime security tasks is reflected in procurement. This is dominated by patrol and logistic support vessels. Prominent amongst the former are the *Tenochtitlan* class inshore patrol vessels, built in the naval ASTIMAR 1 shipyard at Tampico to the Dutch Damen Stan Patrol 4207 design. A contract for a further three vessels in January 2016 takes total orders to ten of the class out of a planned total of up to twenty units. Good progress is also being made with renewed construction of the much larger *Oaxaca* class offshore patrol vessels. *Chiapas*, the first of the new batch – and fifth member of the overall class – was launched on 13 November 2015. Work has also started on a sixth ship, to be named *Hidalgo.*

The elderly – and often somewhat decrepit – state of the handful of former US Navy vessels that form the small fleet of front-line surface escorts is however leading to plans for replacement of these ships. It seems that the potential transfer of retired FFG-7 class frigates from the United States has been rejected in favour of domestic construction of a new light frigate design. A variant of the Damen Sigma design would seem a likely front-runner given the track record of successful co-operation with the Dutch company. However, the French 'Gowind' type may also be considered. It appears that a total of up to six ships is ultimately envisaged to replace existing *Bronstein* (FF-1037) and *Knox* (FF-1052) class frigates on a like-for-like basis. It is hoped the first ship will be in service around 2020.

Elsewhere in the region, the recapitalisation of the Royal **Bahamas** Defence Force under the US$230m 'Sandy Bottom' project is now close to completion. Two Damen Stan Patrol 3007 vessels and a Stan Lander 5612 landing craft were commissioned in March 2016. Four Stan Patrol 4207 patrol ships had been delivered earlier and the nine-ship programme will be completed with the arrival of two further Stan Patrol 3007 types before the end of 2016. The programme also includes the refit of two older boats and a major renewal of harbour infrastructure. All of this will be completed in the next twelve months, after which attention may turn to improving aerial surveillance capabilities.[18]

MAJOR SOUTH AMERICAN NAVIES - BRAZIL

Brazil is currently in the middle of a complicated economic and political crisis. 2016 is set to see the first multi-year recession since 1930–1 on the back of the sharp decline in commodity prices and a corruption scandal that has choked-off construction activity and indirectly resulted in the temporary suspension of President Dilma Rousseff whilst she faces impeachment on charges of financial irresponsibility and other irregularities.[19] Underlying the political crisis are charges of corruption that implicate much of Brazil's political class and many of its leading companies.

Prominent amongst the companies affected is the giant Odebrecht conglomerate. Its ICN subsidiary is contracted to complete the giant new submarine manufacturing complex and base facility at Itaguaí in conjunction with France's DCNS under the

The Mexican Navy's *Oaxaca* class offshore patrol vessel *Independencia* seen in 2012. The tug gives some idea of the diminutive nature of these ships. Four were delivered between 2003 and 2010 and another two, with different armament and other variations, are currently under construction. *(US Navy)*

Table 2.1.6: BRAZILIAN NAVY: PRINCIPAL UNITS AS AT MID 2016

TYPE	CLASS	NUMBER	TONNAGE	DIMENSIONS	PROPULSION	CREW	DATE
Aircraft Carriers							
Aircraft Carrier – CV	**SÃO PAULO** (FOCH)	1	33,500 tons	265m x 32/51m x 9m	Steam, 30 knots	1,700	1963
Principal Surface Escorts							
Frigate – FFG	**GREENHALGH** (Batch I Type 22)	2	4,700 tons	131m x 15m x 4m	COGOG, 30 knots	270	1979
Frigate – FFG	**NITERÓI**	6	3,700 tons	129m x 14m x 4m	CODOG, 30 knots	220	1976
Corvette – FSG	**BARROSO**	1	2,400 tons	103m x 11m x 4m	CODOG, 30 knots	145	2008
Corvette – FSG	**INHAÚMA**	3	2,100 tons	96m x 11m x 4m	CODOG, 27 knots	120	1989
Submarines							
Submarine – SSK	**TIKUNA** (Type 209 – modified)	1	1,600 tons	62m x 6m x 6m	Diesel-electric, 22 knots	40	2005
Submarine – SSK	**TUPI** (Type 209)	4	1,500 tons	61m x 6m x 6m	Diesel-electric, 22+ knots	30	1989
Major Amphibious Units							
Landing Ship Dock – LSD	**BAHIA** (FOUDRE)	1	12,000 tons	168m x 24m x 5m	Diesel, 20 knots	160	1998

PROSUB programme. Whilst the main elements of construction work are continuing, progress has been significantly slowed due to sharp budgetary reductions. Priority is being given to those parts of the complex required to support submarine assembly, which is also well underway. However, in June 2016, it became clear that assembly of the four 'Scorpène' type submarines being delivered under the project has been further delayed. The launch of the first boat, *Riachuelo*, is now scheduled for 2018. Delivery will follow in July 2020, some five years later than first planned. The three other boats will follow between 2021 and 2023. The future of the planned nuclear-powered submarine, *Álvaro Alberto*, the first in Latin America, seems subject to even more uncertainty and the likely start of construction remains uncertain.[20]

If the PROSUB project is beset with difficulties, the challenges facing the surface fleet are even more substantial. Whilst the long-planned PROSUPER renewal programme to acquire new frigates and offshore patrol remains alive, there is little likelihood of funds being found for new surface construction in the near future. In the meantime, the operational fleet is shrinking as older vessels wear out. The frigate *Bosísio* and corvette *Frontin* were both retired as earlier announced in September 2015, the latter subsequently being expended in a live-fire exercise in April 2016. The use of patrol vessels of the *Amazonas* class in place of the usual surface escorts to support the Brazilian Navy's maritime contribution to the United Nations Interim Force in Lebanon (UNIFIL) is one indication of the degree of strain the surface fleet is under. However, construction of smaller patrol ships is also proving problematic. In early 2016, the navy decided to terminate a contract for the construction of follow-on *Macaé* class vessels with shipbuilder Estaleiro Ilha due to significant delays with the project. Assembly of some of the five ships that comprise this second batch – two were previously completed by another yard – is quite well advanced. The intention seems to be to get another company to complete the order.

Other ships being withdrawn over the past twelve months include the *Aratu* class minesweeper *Abrolhos* and the dock landing ship *Ceará*. The latter suffered a major propulsion failure in 2015 and has been replaced by the newly-acquired amphibious transport dock *Bahia*, formerly the French Navy's *Sirocco*. She was commissioned at Toulon in March 2016, arriving in Brazil the following month. Meanwhile, the proposed modernisation of the sole aircraft carrier, *São Paulo*, may be put in jeopardy if reports of substantial cracks in her hull are

The Brazilian Navy's amphibious transport dock *Bahia*, formerly the French Navy's *Sirocco*, pictured off Toulon in March 2016, shortly before her delivery voyage. Her acquisition represents a bright spot in an otherwise rather dismal year for the Brazilian Navy. *(Brazilian Navy)*

confirmed.[21] The 55-year-old ship – the former French *Foch* – has been plagued by unreliability since her acquisition in 2000 and a major, four-year long refit has been planned to enable her to serve into the 2030s. Planned enhancements include a new propulsion system, reconstruction of catapults and lifts to handle heavier aircraft and a new combat system and sensors. Inevitably, this extensive package of work may not be cost-effective if damage is found to be substantial.

OTHER SOUTH AMERICAN NAVIES

Elsewhere in Latin America, the change of presidency in **Argentina** is likely to see new President Mauricio Macri assume a lower profile towards the Falkland Island/Malvinas sovereignty dispute.[22] There seems to be little prospect in any significant uplift in naval investment given the country's economic challenges. Indeed, the re-establishment of an Argentine Air Force interceptor capability is likely to have first call on any available funds following retirement of the country's Dassault Mirage fighters at the end of 2015. There have been various reports of planned purchases of second-hand ships – notably surplus A-69 corvettes from France and amphibious ships from the United States – as well as a re-start of the delayed offshore patrol vessel programme. It seems wise to take these reports with a pinch of salt until firm contracts are evidenced given the lack of past progress and the struggle even to keep a reasonable proportion of the current fleet operational. Previous reports that emerged in early 2015 relating to the potential acquisition of Chinese-built P18 corvettes have not yet been followed by any further developments. Any deal would now seem less likely given the Macri regime's closer links with the United States. The sinking of a Chinese vessel that was supposedly illegally fishing in Argentine waters by the Argentine Coast Guard in March 2016 is also unlikely to have strengthened relations between the two countries.

Neighbouring **Chile** is largely focused on modernisation of existing naval assets. A major modernisation of the three British-built Type 23 frigates is likely to be contracted by the end of 2016. Domestic shipyard ASMAR has been undertaking the first refit of the 'Scorpène' type submarine *O'Higgins* and started work on the fourth *Piloto Pardo* Fassmer OPV-80 type patrol vessel, *Cabo Odger*, in May 2015. She will incorporate the improved armament and ice-strengthening applied to the third member of the class. Delivery is planned for 2017. Future construction plans include two further offshore patrol vessels and a new icebreaker.

Elsewhere in South America, **Peru** continues to make progress with its fleet renewal plan. This includes the local modernisation of its principal surface combatants and submarines, as well as indigenous construction of amphibious ships and patrol vessels. The life extension of its veteran *Lupo* class frigates has taken a major step forward with the imminent return to the fleet of the first two vessels – *Aguirre* and *Bolognesi* – to complete their refits. Enhancements include an upgraded combat system, the installation of a multi-function Kronos radar and Exocet MM40 surface-to-surface missiles.

Notes

1. These figures are simplifications. The US$580bn authorised for FY2016 comprised a base budget of US$522bn and an additional US$58bn for overseas contingency operations (OCO) – essentially the additional cost of fighting wars such as those against the 'Islamic State'. The comparable requests for FY2017 are US$524bn and US$59bn respectively. There has been considerable debate as to whether the OCO 'slush fund' has been partly used to compensate for inadequacies in the core budget as overseas operations have wound down. For further discussion, see Janine Davidson and Emerson Brooking's 'How the Overseas Contingency Operations Fund Works – and Why Congress Wants to Make it Bigger' published on the Council on Foreign Relations website on 16 June 2015. The report's co-author, Janine Davidson, assumed the position of Under Secretary of the US Navy in March 2016.

2. See further the *Fiscal Year 2017 Budget Request Briefing* (Washington DC: Office of the Under Secretary of Defense (Comptroller)/CFO, 2016). Full budget briefing materials can be found at: http://comptroller.defense.gov/Budget Materials.aspx.

3. The figures approved for FY2016 comprised US$159.3bn in base funding and US$9.4bn for OCO. The respective figures proposed for FY2017 are US$155.4bn and US$9.5bn, resulting in an overall decline of 2.3 per cent.

4. The potential increase in the cost of the Canadian Surface Combatant programme was revealed in James Cudmore's 'Cost to build navy's new warships more than doubles to $30bn', in a CBC News report of 1 December 2015. The report stated that the cost analysis was carried out independently by international consultancy A. T. Kearney. It suggested that total programme expenses could rise to C$42bn (compared with an original budget of C$26.2bn) when provision of ammunition, infrastructure and support services was taken into account. A copy can currently be found at: www.cbc.ca/news/politics/nsps-naval-ship-procurement-costs-1.3345435.

5. This story has also been covered by CBC News' James Cudmore in 'Davie Shipyard's $700m deal for navy supply ship retrofit to go ahead'. The story is currently available at: www.cbc.ca/news/politics/davie-shipyard-s-700m-deal-for-navy-supply-ship-retrofit-to-go-ahead-1.3344037. It indicates that CAD$490m of this relates to the supply of the ship and its services – including a core crew. The balance encompasses fuel, port services and other marine costs. The navy will provide a small number of personnel to support refuelling and other naval activities.

6. Under the Canadian National Shipbuilding Strategy, Irving's Halifax Shipyard has been allocated the construction of war-fighting vessels. Seaspan's Vancouver Shipyards has been given responsibility for naval auxiliaries and other major government-owned ships.

7. Much of the information contained in this paragraph is based on the *Report to Congress on the Annual Long-Range Plan for Construction of Naval Vessels for FY2017* (Washington DC: US Navy, 2016). Ronald O'Rourke's periodically updated *Navy Force Structure and Shipbuilding Plans: Background and Issues for Congress RL32665* (Washington DC: Congressional Research Service) and available on the FAS website www.fas.org/sgp/crs/ provides a good explanation of the various trends and issues arising from these annual reports.

8. For further detail, see Megan Eckstein's 'Navy Revising Force Structure Assessment In Light Of Increased Attack Sub, Other Ship Needs', *USNI News* – 29 February 2016 (Annapolis MD: US Naval Institute, 2016). A copy can currently be found by searching www.news.usni.org/.

9. The teething troubles and cost overruns associated with AAG have been covered in a series of reports by Megan Eckstein and Sam LaGrone in *USNI News*, *ibid*.

10. Amongst a number of journalists reporting on this issue was Tom Capaccio in 'General Dynamics Destroyer Reviewed by U.S. for Cancellation', *Bloomberg* – 14 September 2015 (New York: Bloomberg LP, 2015).

11. It was previously envisaged that eleven additional Flight IIA destroyers would be ordered. Construction would then shift to the Flight III design with effect from DDG-124, the second ship planned in the FY2016 programme. However, Congress authorised part-funding for an

The SIMA Shipyard in Callao is currently contracted to modernise four of the class. A further two vessels are expected to join the programme in due course. SIMA is also to undertake a refit of the navy's four youngest Type 209 submarines – *Angamos*, *Antofagasta*, *Pisagua* and *Chipana* – with the help of ThyssenKrupp Marine Systems and Israel's Elbit Systems. A significant upgrade programme is underway at the yard prior to the start of the project in 2017. Another SIMA project is for *Paita*, the first of two *Makassar* type amphibious transport docks, which should be launched around the end of 2016. SIMA has already delivered the sail training ship *Unión*, which was commissioned on 27 January 2016.

In addition to its salt-water commitments, the *Marina de Guerra del Perú* also has important responsibilities policing the Peruvian Amazon. *Jane's Defence Weekly* shed new light on this operation when they interviewed the head of the SIMA Shipyard in Iquitos in January 2016.[23] The facility has the capability of constructing shallow-draft vessels of up to 1,500 tons displacement. It is currently in the course of delivering 350-ton *Manuel Clavero* class river gunboats, CB-90 derived *Rio Itaya* class interceptors and an innovative class of humanitarian support vessels – the itinerant platforms for sustained social action (PIASS) – that provide a range of support services to isolated riverine communities. The navy has also previously taken delivery of a fleet of Griffon hovercraft to counter narcotics trafficking in the area.

Colombia has also been developing its indigenous naval construction capabilities. Shipbuilder COTECMAR is currently completing a third Fassmer-designed OPV-80 patrol vessel and a series of 6,000-ton landing craft. At least one landing craft is also to be built for Honduras. The second OPV-80 type vessel, *7 de Augusto*, carried out a successful deployment to the Indian Ocean in the second half of 2015, operating with European Union and NATO forces on anti-piracy duties. The end of the fifty-year war with the revolutionary guerrilla movement FARC, heralded by the signature of a peace agreement in Havana, Cuba on 23 June 2016, may result in greater future involvement in international stabilisation missions of this kind. In the meantime, the fleet's underwater capabilities have been strengthened by the long-awaited arrival of the two former German Navy Type 206 submarines acquired in August 2012. They have now completed lengthy refits – including modifications for warm-water operation – at ThyssenKrupp Marine Systems' HDW yard in Kiel and were commissioned at the end of 2015.[24] Meanwhile, **Uruguay** is also reportedly considering purchases of OPV-80 type vessels, although second-hand alternatives are also being considered.

The economic crisis in **Venezuela** continues to impact both modernisation plans and operational availability. For example, repairs to the POVZEE class offshore patrol vessel *Warao*, heavily damaged in a grounding incident in 2013, appear to have ground to a halt for lack of funds. Her sister *Guaiqueri* may also need repairs after an explosion in March 2016 that left one crew member dead and several others badly injured.

---

additional, third destroyer in the FY2016 budget. The opportunity has therefore been taken to squeeze another Flight II ship into planned construction. As such, it will be the third FY2016 ship, the future DDG-125, that will be the lead Flight III variant. Interestingly, however, the part funded-ship does not appear in the official shipbuilding plan. The allocation of this additional ship to BIW reflects a 2002 agreement re-assigning DDG-51 and LPD-17 amphibious transport dock construction between BIW and what is now HHI. As Congress had recently increased construction of LPD-17 type ships – built exclusively by HHI – by one, BIW was effectively 'owed' an additional destroyer.

12. Ronald O'Rourke's *Navy Littoral Combat Ship (LCS)/Frigate Program: Background and Issues for Congress RL33741* (Washington DC: Congressional Research Service), also available on the FAS website, is probably the best source of information available on the LCS programme's convoluted history. The prospect of the interchangeable mission module approach being overturned is discussed more fully by David Larter in 'Navy weighs radical changes to littoral combat ship crews, module' in the *Navy Times* – 14 June 2016 (Springfield VA: Sightline Media Group, 2016).

13. The detailed development of the class was covered by Scott Truver in 'The US Navy's LPD-17 Class: Snake Bit No More!', *Seaforth World Naval Review 2016* (Barnsley: Seaforth Publishing, 2015), pp.104–23.

14. The new designations – agreed by Secretary of the Navy Ray Mabus in conjunction with the then Chief of Navy Operations Jonathan Greenert in September 2015 – introduces a new root designator – E for expeditionary support – to US Navy classifications. See Chris Cavas' 'A New Class of Ship – 'Expeditionary Support' *Navy Times* – 4 September 2015 (Springfield VA: Sightline Media Group, 2015). Analysis of the ships covered by the new designations has also been provided by Scott Truver in previous editions of *World Naval Review*.

15. Full detail on the *Virginia* class programme can be found in another of Ronald O'Rourke's reports, viz. *Navy Virginia (SSN-774) Class Attack Submarine Procurement: Background and Issues for Congress RL32418* (Washington DC: Congressional Research Service).

16. This story was covered extensively by David Larter in an article entitled 'Secret weld: How shoddy parts disabled a $2.7 billion submarine', the *Navy Times* – 28 March 2016 (Springfield VA: Sightline Media Group, 2016).

17. For further detail, see Sydney J. Freedberg Jnr.'s 'Ohio Replacement Plan is Good News for Electric Boat', *Breaking Defence* – 29 March 2016 (New York: Breaking Media Inc., 2016).

18. Progress with the upgrade programme was covered in greater detail by Alejandro Sanchez in 'Bahamas' Sandy Bottom project upgrades defence capabilities', *Jane's Defence Weekly* – 1 June 2016 (Coulsdon: IHS Jane's, 2016), p.13.

19. Brazil's financial and political woes are far beyond the scope of this annual. For readers requiring more background, a good starting point is Peggy McInerny's 'What's behind Brazil's political and economic crises?' posted to the *UCLA* Newsroom on 16 May 2016 and currently available at: newsroom.ucla.edu/stories/what-s-behind-brazil-s-economic-and-political-crises.

20. The Portuguese language Poder Naval website at www.naval.com.br/blog/ remains an excellent source of information on current Brazilian Navy developments.

21. See Victor Barreira, '"Cracks found" in Brazil's aircraft carrier', *Jane's Defence Weekly* – 11 November 2015 (Coulsdon: IHS Jane's, 2015), p.10.

22. The Spanish-language websites infodefensa.com and defensa.com have been useful sources of information for this section.

23. See further Guy Toremans' interview with Captain Luis Diaz Ramirez in *Jane's Defence Weekly* – 3 February 2016 (Coulsdon: IHS Jane's, 2015), p.34.

24. Further details were provided in Eric Wertheim's 'Combat Fleets', *Proceedings* – vol. 142/4, April 2016 (Annapolis MD: US Naval Institute, 2016), p.92.

2.2 REGIONAL REVIEW

ASIA AND THE PACIFIC

Author:
Conrad Waters

INTRODUCTION

China's ongoing creation and militarisation of islands from the reefs of the South China Sea has remained the dominant influence on Asian naval developments over the last twelve months. The massive land reclamation strategy – which encompasses seven new islands in the Spratly group to date and has become known as the 'great wall of sand' – is aimed at solidifying China's contentious 'nine dash line' claim to much of the area. Around 3,000 acres (nearly five square miles) of new territory has been created and equipped with piers, harbours, support facilities and – in some cases – military airstrips.[1] To date, it is the Philippines that has been most impacted by the island-building strategy. Its arbitration case against China's claims at the Permanent Court of Arbitration in the Hague – a case in which China has refused to participate – is expected to be determined by mid-July 2016. Most commentators expect the result will have little impact on future Chinese actions.

The major regional concern is what will happen next. There have already been signs that China intends to replicate its actions in the Spratly Islands in the Paracel group, which is the subject of overlapping claims with Vietnam.[2] Incidents have also occurred further afield. In June 2016, an Indonesian 'Parchim I' class corvette opened fire on a group of Chinese fishing vessels in the disputed Natuna Islands region after they failed to obey its instructions. The action brought swift Chinese protests for an abuse of force in Beijing's 'traditional fishing grounds'. To the north, there have also been further incidents near the Senkaku/Diaoyutai group in the East China Sea, where the level of incursions by Chinese vessels into waters claimed by Japan has been rising markedly.

It is noteworthy that China appears willing to press its claims so assertively given the increasingly strident condemnation that they bring. As indicated in this book's Introduction, China's stance may partly be driven by a calculation that it is unlikely to face more than token and fragmented opposition to its actions given the scale of its economic – and military – power. Certainly, Chinese global economic influence remains on the rise – its revival of the ancient Silk Road between Asian and Europe under the 'One Belt, One Road' (OBOR) policy is channelling hundreds of billions of Chinese investment into much-needed infrastructure that tends to constrain criticism of its less constructive international activities. For some, China's economic diplomacy may be the 'only game in town' worth following.[3] It may also be that it perceives its creation of 'facts on the ground' can only strengthen its position if it eventually decides to come to the negotiating table.

Nevertheless, there is evidence that tangible opposition to Chinese activity is continuing to harden. Much of this is being driven by the United States and its allies, motivated by concerns with respect to both rights to freedom of navigation and the wider 'rules based' international order. A potentially significant development was the US Navy's deployment of the destroyer *Lassen* (DDG-82) to transit waters off the Subi Reef, one of the new islands, in October 2015 in the first of a planned series of freedom of navigation exercises in the region. This action is being reinforced by other activities, including the periodic deployment of the new Poseidon P8-A maritime patrol aircraft to Singapore in addition to a number of existing bases. Indeed, increased American aerial activity in the region from several bases has resulted in a number of interceptions by Chinse planes.

The network of alliances between countries troubled by China's assertive action – referenced in last year's *Seaforth World Naval Review* – is also continuing to strengthen. Perhaps most significantly, Japan was invited to become a permanent participant in the annual Indo-American Malabar series of exercises from October 2015. The development was lauded in the Indian press as indicating '… a new strategic realignment in the Indo Pacific region'.[4] There has also been tangible progress with the United States' 2014 Enhanced Defense Cooperation Agreement (EDCA) with the Philippines. The deal giving American access to five additional base areas rebuilds the relationship to a level not seen since the Subic Bay and Clark Field bases closed in the early 1990s. Nevertheless, the United States has been keen to stress that it does not support any particular sovereignty claims amongst the various disputes and still hopes that China can still be persuaded '… to be part of the system of cooperative nations that have made the Asian miracle possible'.[5] One practical demonstration of this has been the People's Liberation Army Navy (PLAN) participation in the RIMPAC 2016 exercises off Hawaii – the world's largest – for only the second time.

The Chinese PLAN destroyer *Xian* approaching Pearl Harbor on 29 June 2016 to participate in the US-sponsored RIMPAC 2016 exercises. The United States continues to encourage Chinese participation in such collaborative engagement with other Asian navies in spite of tension caused by its island-building activities in the South China Sea and elsewhere. *Xian*, the sixth and final member of the Type 052C class, was commissioned in February 2015. PLAN has now switched to the improved Type 052D variant, which incorporates an improved multi-function radar and a revised vertical launch system (VLS). *(US Navy)*

Table 2.2.1: FLEET STRENGTHS IN ASIA AND THE PACIFIC – LARGER NAVIES (MID 2016)

COUNTRY	AUSTRALIA	CHINA	INDONESIA	JAPAN	S KOREA	SINGAPORE	TAIWAN	THAILAND
Aircraft Carrier (CV)	–	1	–	–	–	–	–	–
Support/Helicopter Carrier (CVS/CVH)	–	–	–	3	–	–	–	1
Strategic Missile Submarine (SSBN)	–	6	–	–	–	–	–	–
Attack Submarine (SSN)	–	8	–	–	–	–	–	–
Patrol Submarine (SSK/SS)	6	50	2	17	14	4	4	–
Fleet Escort (DDG/FFG)	11	60	6	37	24	6	24	8
Patrol Escort/Corvette (FFG/FSG/FS)	–	35	25	6	18	6	1	11
Missile Armed Attack Craft (PGG/PTG)	–	75	20	6	17	–	c.30	6
Mine Countermeasures Vessel (MCMV)	6	25	11	27	9	4	10	6
Major Amphibious Units (LHD/LPD/LSD)	3	4	5	3	1	4	1	1

Notes: Chinese numbers approximate; Some additional Indonesian patrol gunboats are able to ship missiles; Taiwan's submarines are reported to have limited operational availability.

MAJOR REGIONAL POWERS – AUSTRALIA

Australia's 2016 defence white paper, published on 25 February after delays due to prime ministerial and other governmental changes, set a clear maritime slant to procurement priorities. Confirming previous plans to ensure a sustainable Australian naval shipbuilding sector, the white paper reaffirmed a number of previously-announced naval construction programmes and announced additional investment in maritime capabilities. In contrast to many other government announcements the world over, the white paper was quickly followed by decisions intended to make these plans a reality.

The plans outlined in the white paper covered a wide range of different naval vessels and equipment. These were further detailed in an integrated investment programme and are summarised below:[6]

Submarines: The long-standing plan to replace the existing six *Collins* class submarines with twelve new boats under Project SEA 1000 was confirmed. This resolved previous concerns that the hugely-expensive programme might be reduced to just eight units. The project, which is estimated to cost c. AU$50bn (US$38bn) and will be the largest defence programme in Australia's history, is to be carried out on a rolling basis. Under this, the first of the new submarines will begin entering service in the 2030s and be approaching retirement as the last is delivered around 2050.

On 26 April 2016 it was announced that DCNS' 'Shortfin Barracuda' design had won out over rival bids from Germany's Thyssen Krupp Marine Systems (TKMS) and a Japanese consortium led by Mitsubishi Heavy Industries (MHI). This overturned previous assumptions that Japan's *Soryu* class was the front runner for the contract, probably reflecting a loss of political support after former Australian Prime Minister Tony Abbott – one of the Japanese design's main cheerleaders – was removed from office in September 2015. The selected design uses the *Suffren* class nuclear-propelled attack submarines being built for the *Marine Nationale* as its design reference. However, the propulsion system will owe much to DCNS' smaller conventional submarines, including the successful 'Scorpène'.[7] Overall length is reported as being 97m, whilst displacement is in the region of 4,700 tons. The April announcement confirmed that all the submarines would be constructed at Adelaide, presumably at the ASC shipyard on which Australia's future shipbuilding industry is being concentrated.

Surface Combatants: In line with plans announced in August 2015, the defence white paper confirmed that construction of a new class of frigates under Project SEA 5000 would commence in Adelaide in 2020. The project will follow-on from construction of the *Hobart* class air-defence destroyers and will cost over AU$30bn (US$23bn). It will see the eight existing *Anzac* class ships replaced by nine new vessels from the late 2020s onwards. This will increase the overall size of the surface fleet from eleven to twelve ships. In April 2016 it was announced that the British Type 26 and Italian FREMM frigates had been shortlisted for the design contract alongside a modified version of the Spanish F-100 class hull used for the *Hobart* class. The new ships will be optimised for anti-submarine warfare. This suggests the Spanish air-defence design will need significant adjustment if it is to match the capabilities of its two competitors. A decision on the successful bidder will be taken by 2018.

Patrol Vessels: A gap in production between the end of *Hobart* class construction and the ramp-up of work on the new frigates has also led to acceleration of Project SEA 1180 for new offshore patrol vessels. The white paper confirmed that work on these will start in 2018 and stated that twelve would replace the remaining thirteen members of the *Armidale* class. At between 70m and 80m in length, they will be larger than the existing ships. Designs from the Dutch Damen group and from Germany's Fassmer and Lürssen have been shortlisted for the contract. Whilst Adelaide will build the initial vessels, it is envisaged that production will shift to Western Australia once work on the new frigates gets underway. It was initially envisaged that the Project SEA 1180 design would also replace existing mine countermeasures and hydrographic vessels, leading to a larger programme. However, the white paper

A computer-generated image of the DCNS 'Shortfin Barracuda' design, which won out over German and Japanese competition to be selected for Australia's *Collins* class replacement programme. The programme – Project SEA 1000 – involves the acquisition of twelve boats for delivery from the early 2030s through to 2050 at an estimated cost of some AU$50bn. *(DCNS)*

Navantia is one of three shipbuilders who have been asked to propose designs for the Royal Australian Navy's Project SEA 5000, which envisages the acquisition of nine frigates. The Navantia proposal is very clearly related to its existing F-100 type, which is already being built in Australia as the *Hobart* class air-defence destroyers. Whilst this provides Navantia with the benefits of an incumbent position, competing British and Italian designs are optimised for anti-submarine warfare, which is a key requirement for the new ships. A decision will be made in 2018. *(Navantia)*

Table 2.2.2: ROYAL AUSTRALIAN NAVY: PRINCIPAL UNITS AS AT MID 2016

TYPE	CLASS	NUMBER	TONNAGE	DIMENSIONS	PROPULSION	CREW	DATE
Principal Surface Escorts							
Frigate – FFG	**ADELAIDE** (FFG-7)	3	4,200 tons	138m x 14m x 5m	COGAG, 30 knots	210	1980
Frigate – FFG	**ANZAC**	8	3,600 tons	118m x 15m x 4m	CODOG, 28 knots	175	1996
Submarines							
Submarine – SSK	**COLLINS**	6	3,400 tons	78m x 8m x 7m	Diesel-electric, 20 knots	45	1996
Major Amphibious Units							
Amph Assault Ship – LHD	**CANBERRA** (JUAN CARLOS I)	2	27,100 tons	231m x 32m x 7m	IEP, 21 knots	290	2014
Landing Ship Dock – LSD	**CHOULES** (LARGS BAY)	1	16,200 tons	176m x 26m x 6m	Diesel-electric, 18 knots	160	2006

The RAN's Plan Pelorus emphasises integrated task groups as opposed to the previous practice of single ship deployments. Key to this concept are the new *Canberra* class LHDs and the *Hobart* class air-defence destroyers. These images show (top) *Canberra* operating with RAN and RNZN frigates and (bottom) *Hobart* in the course of outfitting. *(RAN/AWD Alliance)*

stated that the service lives of four of the existing *Huon* class minehunters would be extended into the 2030s. Modular equipment embarked on non-specialised vessels might be used to fulfil the hydrographic role.

Auxiliaries: The other major programme confirmed in the defence white paper was Project SEA 1654 Phase 3 for two replenishment ships. Unlike the other warships, capacity issues and lack of suitable infrastructure means that these ships will be built abroad. In March 2016 it was announced that a variant of Spanish group Navantia's *Cantabria* had been selected as preferred bidder for the contract, beating a rival proposal from Daewoo Shipbuilding & Marine Engineering (DSME) based on a BMT Aegir 18A design. The white paper has also indicated that a third replenishment vessel – or an additional logistics support ship similar to the existing *Choules* – will be ordered at a later stage.

Whilst the capabilities provided by this long list of projects will take time to deliver, the Royal Australian Navy is also starting to see the benefits of previous defence acquisitions. Both new *Canberra* class amphibious assault ships are now in service following the commissioning of *Adelaide* on 4 December 2015. They will form a key part of the future naval strategy outlined in the April 2015 Plan Pelorus. This switches emphasis from deploying independent ships towards a focus on the inter-linked capabilities provided by self-contained task groups. Another key element of these groups will be provided by the new *Hobart* class. The much-delayed construction of these is now well advanced and the lead ship should commence sea trials in the second half of 2016. The second unit, *Brisbane*, will also be launched within the next twelve months and the keel of the third and final ship, *Sydney*, was formally laid on 19 November 2015. Her namesake, the FFG-7 type frigate *Sydney* (FFG-03), had been decommissioned some two weeks previously after spending the last few months of her operational life as an alongside training ship at Fleet Base East Sydney.

Table 2.2.2 summarises Royal Australian Navy fleet composition as at mid-2016.

MAJOR REGIONAL POWERS – CHINA

China's PLAN has steadily undertaken a course of qualitative development under which the process of replacing large numbers of technological-obsolete warship designs from the Cold War era by a fleet composed of increasingly standardised ships and equipment is now largely complete. At the same time, work is continuing on new weapons for example aircraft carriers, nuclear submarines and advanced anti-ship missiles – which would be essential in deterring or defeating a peer competitor should current tensions unexpectedly spill over into armed conflict.

Prominent amongst these new systems are China's anti-ship ballistic missiles. A major military parade in Beijing in September 2015 to mark the seventieth anniversary of the Second World War provided an opportunity to showcase the first of these, the Dong Feng DF-21D. An anti-ship variant of the longer-range DF-26 – nicknamed the 'Guam Express' by virtue of its ability to threaten US bases on that island – has reportedly also been developed. If true, the missile would potentially threaten US Navy carrier groups operating east of Taiwan and the Philippines, significantly limiting their operational room for manoeuvre. China, of course, is developing its own fleet of aircraft carriers and there has been some speculation as to how these might be used. An interesting insight was provided by *Jane's Defence Weekly* in March 2016. Rather than being used offensively in the fashion of US Navy carrier groups, it was argued that the new ships would primarily be deployed defensively, for example strengthening air defences in areas not covered by China's artificial islands in the South China Sea.[8]

Table 2.2.3 highlights the PLAN's most important warship classes. Additional detail on recent developments with respect to key types is provided below.[9]

An image of five different classes of PLAN surface vessel alongside at Pearl Harbor at the start of the RIMPAC 2016 exercise. Furthest from the camera are the Type 054A frigate *Heng Shui* – commissioned in 2012 – and the Type 052C class destroyer *Xian* (2015). These currently represent the PLAN's most numerous blue-water surface combatants. Closer to the camera are the Type 920 hospital ship *Peace Ark* (2008), the Type 903A replenishment ship *Gaoyouhu* (2016) and the Type 926 submarine rescue ship *Changdao* (2012). Their deployment represents the PLAN's increasingly heavy investment in support vessels to sustain blue water operations. It is noteworthy that the average age of this substantial flotilla is less than four years. *(US Department of Defense)*

Aircraft Carriers and Amphibious Ships: Official confirmation that construction of China's first indigenous aircraft carrier is underway was provided at the end of 2015. Previous unofficial reports had suggested work on the Type 001A class vessel commenced at Dalian shipyard towards the end of 2013 and satellite images have tracked the course of its construction from March 2015 onwards. The new ship has similar dimensions to the existing Short Take-Off But Arrested Recovery (STOBAR) type *Kuznetsov* class carrier *Liaoning* and government sources have confirmed she will be an improved variant of the existing Soviet-built ship. This practical approach should facilitate the construction and entry into service of the new ship, which commentators believe will be ready for launch during 2017.

Meanwhile, a fourth Type 071 'Yuzhao' class amphibious transport dock, *Yimeng Shan*, was commissioned into service with the East Sea Fleet on 1 February 2016. The other three ships in the class have all been deployed with the South Sea Fleet. Commentators suggest a total of between six and nine ships are ultimately envisaged to provide a core amphibious capability to each fleet. The East Sea fleet has also been the leading beneficiary of renewed deliveries of the smaller Type 072A landing ships, production of which is now estimated to extend to around fifteen vessels. Chinese media reports suggest that three members of the class were inducted into service in a single day on 7 March 2016.

Surface Combatants: Chinese surface combatant construction continues to be focused on series production of standardised destroyer, frigate and corvette designs. Between three and four of the Type 052D 'Luyang III' destroyer class are now in commission and at least five more have been launched of a rumoured twelve-ship class. Members of the earlier six-strong Type 052C 'Luyang II' batch are being increasingly seen on extended international deployments, demonstrating confidence in the basic reliability of the design. Deliveries of these modern ships are allowing ongoing withdrawals of the obsolete 1970s-generation Type 051 'Luda' class, no more than a half-dozen of which now remain in service. Other interim Chinese destroyer types, as well as the Russian-built *Sovremenny* class, are being upgraded with the latest indigenous weapons and sensors.

Production of the Type 054A 'Jiangkai II' frigate class now extends to at least twenty-five units, of which twenty-two are in commission. The design's evident success means that it now dominates the PLAN's general-purpose frigate force and many older vessels are either being retired or transferred to second-line duties. Notably, all four members of the Type 053H2G 'Jiangwei I' type have been largely disarmed and converted to coast guard use over the last twelve months. Interestingly, the basic Type 054A design has also been adopted to form a new class of coast guard patrol vessel, surrendering the frigate type's missiles and associated sensors but retaining its long-range propulsion system and helicopter-handling facilities. The move follows the precedent set with the smaller Type 056 'Jingdao' class corvette design, which has also been adapted for a series of smaller coast guard patrol vessels.[10]

The PLAN Type 054A frigate *Yiyang* pictured in the course of a visit to the US Navy's Atlantic base at Mayport, Florida in November 2015. Twenty-two of the class are currently in commission and they form the mainstay of the PLAN's international deployments. *(US Navy)*

Table 2.2.3: PEOPLE'S LIBERATION ARMY NAVY: PRINCIPAL UNITS AS AT MID 2016

TYPE	CLASS	NUMBER	TONNAGE	DIMENSIONS	PROPULSION	CREW	DATE
Aircraft Carriers							
Aircraft Carrier – CV	Project 1143.5/6 **LIAONING** (Kuznetsov)	1	60,000 tons	306m x 35/73m x 10m	Steam, 32 knots	Unknown	2012
Principal Surface Escorts							
Destroyer – DDG	Type 052D **KUNMING** ('Luyang III')	3	7,500 tons	156m x 17m x 6m	CODOG, 28 knots	280	2014
Destroyer – DDG	Type 051C **SHENYANG** ('Luzhou')	2	7,100 tons	155m x 17m x 6m	Steam, 29 knots	Unknown	2006
Destroyer – DDG	Type 052C **LANZHOU** ('Luyang II')	6	7,000 tons	154m x 17m x 6m	CODOG, 28 knots	280	2004
Destroyer – DDG	Type 052B **GUANGZHOU** ('Luyang I')	2	6,500 tons	154m x 17m x 6m	CODOG, 29 knots	280	2004
Destroyer – DDG	Project 956E/EM **HANGZHOU** (Sovremenny)	4	8,000 tons	156m x 17m x 6m	Steam, 32 knots	300	1999
Destroyer – DDG	Type 051B **SHENZHEN** ('Luhai')	1	6,000 tons	154m x 16m x 6m	Steam, 31 knots	250	1998
Destroyer – DDG	Type 052 **HARBIN** ('Luhu')	2	4,800 tons	143m x 15m x 5m	CODOG, 31 knots	260	1994
Plus c. 5 additional obsolescent destroyers of Type 051 **JINAN** ('Luda') class							
Frigate – FFG	Type 054A **XUZHOU** ('Jiangkai II')	22	4,100 tons	132m x 15m x 5m	CODAD, 28 knots	190	2008
Frigate – FFG	Type 054 **MA'ANSHAN** ('Jiangkai I')	2	4,000 tons	132m x 15m x 5m	CODAD, 28 knots	190	2005
Frigate – FFG	Type 053 H3 **LIANYUNGANG** ('Jiangwei II')	10	2,500 tons	112m x 12m x 5m	CODAD, 27 knots	170	1992
Frigate – FSG	Type 056 **BENGBU** ('Jiangdao')	26+	1,500 tons	89m x 12m x 4m	CODAD, 28 knots	60	2013
Plus c.10 additional obsolescent frigates of Type 053 H1/H1G/H2 **TAIZHOU** ('Jianghu II–V') classes							
Submarines							
Submarine – SSBN	Type 094 ('Jin')	c.5	9,000 tons	133m x 11m x 8m	Nuclear, 20+ knots	Unknown	2008
Submarine – SSBN	Type 092 ('Xia')	1	6,500 tons	120m x 10m x 8m	Nuclear, 22 knots	140	1987
Submarine – SSN	Type 093/093G ('Shang')	c.5	6,000 tons	107m x 11m x 8m	Nuclear, 30 knots	100	2006
Submarine – SSN	Type 091 ('Han')	3	5,500 tons	106m x 10m x 7m	Nuclear, 25 knots	75	1974
Submarine – SSK	Type 039A/039B (Type 041 'Yuan')	12+	2,500 tons	75m x 8m x 5m	AIP, 20+ knots	Unknown	2006
Submarine – SSK	Type 039/039G ('Song')	13	2,300 tons	75m x 8m x 5m	Diesel-electric, 22 knots	60	1999
Submarine – SSK	Project 877 EKM/636 ('Kilo')	12	3,000 tons	73m x 10m x 7m	Diesel-electric, 20 knots	55	1995
Plus c.15 obsolescent patrol submarines of the Type 035 ('Ming' class). A Type 032 'Qing' trials submarine has also been commissioned.							
Major Amphibious Units							
Landing Platform Dock	Type 071 **KULUN SHAN** ('Yuzhao')	4	18,000 tons	210m x 27m x 7m	CODAD, 20 knots	Unknown	2007

The Type 056 and its anti-submarine Type 056A development also appear to have been a success. There are now twenty-six of the class in service following the commissioning of *Qujing* in early June, an impressive total given it is only a little over three years since the first commissioned. At least seven more have been launched to date and construction continues at four separate shipyards. Derivatives of the design have also gained export contracts from Bangladesh and Nigeria.

Submarines: News of PLAN submarine development continues to be shrouded in secrecy. However, a number of reports of varying clarity have emerged over the last year to provide some insight into the current status of key programmes. The most significant strategic project is that for Type 094 'Jin' class strategic submarines. Between four and five have been commissioned since 2007. Each submarine is armed with twelve JL-2 (CSS-NX-14) ballistic missiles, theoretically providing a credible second-strike missile capability. However, there is some doubt amongst experts as to whether any of them has ever conducted a nuclear deterrence patrol. However, recent American intelligence reports increasingly concur that the programme is now sufficiently advanced to make this imminent, if it has not already happened.[11]

There is more tangible news with respect to construction of a second batch of Type 093 'Shang' class nuclear-powered attack submarines to join the two commissioned in 2006–07. Chinese media sources correspond with American intelligence assessments in agreeing that there are four submarines in this batch but there has been disagreement between analysts as to how they will differ from the initial pair. However, the release of recent images showing a distinct hump behind the fin tend to lend credibility to those arguing that the new boats have been modified to house vertically launched cruise missiles. This would accord with the Type 093G (for guided missile) designation frequently assigned to the batch.

There continues to be considerable overseas interest in Chinese conventional submarine technology. For example, Pakistan is reported to have ordered eight of the 'S20/S26' export variant of the Type 039A/Type 041 'Yuan' class and Thailand has also been in on/off negotiations for similar boats.

The JMSDF continues to commission and order new *Soryu* class submarines, such as this unidentified member of the class. The seventh member of the series, *Jinryu*, was commissioned during the course of the year, increasing Japan's submarine flotilla to seventeen operational boats and two training units. *(Japan Maritime Self Defence Force)*

Other Vessels: A lack of replenishment capacity has previously been something of an Achilles Heel in the PLAN's ability to support its increasing international deployments but expanded construction has been rapidly making good this deficiency. Three additional Type 903A replenishment ships have been put into service over the last year. The imminent completion of a further vessel will take the combined Type 903/903A class to a total of eight vessels. Production will then likely shift to a much larger, Type 901 design. This reportedly displaces around double that of the earlier ships at c. 45,000 tons and may well be optimised to support the new aircraft carriers. This is reflected in a reported maximum speed in the order of 25 knots. A prototype unit was launched from Guangzhou Shipyard in December 2015 and is expected to enter service in 2017.

Mine countermeasures capabilities are another area where the PLAN's development has been slower than elsewhere. Modern minehunting vessels are entering service only slowly, with current construction apparently focusing on small numbers of Type 082 II vessels. Reports suggest that just four of this class – which operate in conjunction with remotely-operated submersibles – have been delivered since the first was commissioned in 2005. The deficiency is a somewhat surprising one given the possibility of the PLAN becoming involved in amphibious operations and might reflect difficulties in bringing new technology into service.

MAJOR REGIONAL POWERS – JAPAN

Table 2.3.4 indicates that there has been little change to the JMSDF over the last year. The fleet continues to develop along the lines set out in the *National Defense Program Guidelines for FY2014 and Beyond* and the *Medium Term Defense Program (FY2014-2018)*, themselves both influenced by the *National Security Strategy*. The only significant combatant delivered during the course of the year was the seventh *Soryu* class submarine *Jinryu*. She was commissioned on 7 March 2016. In contrast to previous years, her arrival was not balanced by the withdrawal of an older boat. This increased the number of operational submarines to seventeen (plus two training vessels) and marked a tangible step towards the enlarged force level of twenty-two boats that is now targeted. The FY2016 Defence Budget approved construction of a twelfth member of the *Soryu* class, which will incorporate the higher capacity lithium-ion batteries specified for the latest submarines.[12]

Possibly because of the expense associated with constructing the new *Izumo* class helicopter-carrying destroyers, no major surface combatants were included in the FY2011 construction programme. As such, no deliveries took place in 2016 and the composition of the surface flotillas has shrunk by one ship to forty-six (plus three training vessels) as an older *Hatsuyuki* class destroyer was retired. The FY2016 Defence Budget provided funding for the second Aegis-equipped destroyer envisaged in the current five-year medium-term programme. These ships will be enlarged versions of the current *Atago* class and are likely to be powered by combined gas turbine electric and gas turbine (COGLAG) propulsion. This would facilitate the incorporation of directed-energy weapons such as lasers or electromagnetic railguns at a later date. It remains the intention to increase the surface flotillas to a total of fifty-four ships, of which four will be helicopter-carrying destroyers. Funding continued to be allocated to pay for life extensions of existing destroyer classes to allow this target to be reached, although to do this within the lifespan of the current five-year plan would appear challenging

In contrast to the PLAN, the JMSDF has long accorded a high priority to minehunting duties, tending to order a new vessel every one or two years. October 2015 saw the launch of *Awaji* (MSO-304), the first of a class of 700-ton, fibre-reinforced plastic (FRP) mine-countermeasures vessels that will eventually replace the three 1990s-era, wooden-hulled *Yaeyama* (MS-301) class. Japan has been relatively late in adopting new construction methods to minehunter hulls and the *Awaji* class are the only second series of FRP vessels to be built.

An interesting cultural development occurred on 29 February 2016 when Commander Miho Otani was appointed commanding officer of the *Asagiri* (DD-151) class destroyer *Yamagiri* (DD-152). Commander Otani is the first female officer to command a major Japanese warship. Japan follows in the footsteps of the US Navy (1998), British Royal Navy (2012) and French *Marine Nationale* (January 2016) in taking this step.

MAJOR REGIONAL POWERS – SOUTH KOREA

South Korea's naval priorities continue to be balanced between a need to invest in littoral capabilities to counter the asymmetrical threat posed by its northern rival's numerous – if relatively low technology – submarines and fast attack craft and its desire to play a regional, 'blue water' role. Events of the last few years, including the destruction of the corvette *Cheonan* in 2010 and the progress of North Korea's nuclear and ballistic missile programmes, have tended to support a littoral bias. However, the prestige and influence associated with blue water capabilities continue to exert a strong attraction.

The most important littoral warfare programme currently underway is that for the FFX series of frigates. Five of six Batch 1 *Incheon* class ships have now been delivered by local builders HHI and STX, with *Gwangju*, the final unit in this batch, launched on 11 August 2015. Construction is now transitioning to the second FFX-II or *Daegu* class batch. The first of these was launched by DSME on 7 June 2016. This ship is heavier and longer than the initial batch with a full load displacement in the region of 3,600 tons (3,000 tons) and a length of 122m

Table 2.2.4: JAPAN MARITIME SELF-DEFENCE FORCE: PRINCIPAL UNITS AS AT MID 2016

TYPE	CLASS	NUMBER	TONNAGE	DIMENSIONS	PROPULSION	CREW	DATE
Support and Helicopter Carriers							
Helicopter Carrier – DDH	**IZUMO** (DDH-183)	1	27,000 tons	248m x 38m x 7m	COGAG, 30 knots	470	2015
Helicopter Carrier – DDH	**HYUGA** (DDH-181)	2	19,000 tons	197m x 33m x 7m	COGAG, 30 knots	340	2009
Principal Surface Escorts							
Helicopter Destroyer – DDH	**SHIRANE** (DDH-143)	1	7,500 tons	159m x 18m x 5m	Steam, 32 knots	350	1980
Destroyer – DDG	**ATAGO** (DDG-177)	2	10,000 tons	165m x 21m x 6m	COGAG, 30 knots	300	2007
Destroyer – DDG	**KONGOU** (DDG-173)	4	9,500 tons	161m x 21m x 6m	COGAG, 30 knots	300	1993
Destroyer – DDG	**HATAKAZE** (DDG-171)	2	6,300 tons	150m x 16m x 5m	COGAG, 30 knots	260	1986
Destroyer – DD	**AKIZUKI** (DD-115)	4	6,800 tons	151m x 18m x 5m	COGAG, 30 Knots	200	2012
Destroyer – DDG	**TAKANAMI** (DD-110)	5	6,300 tons	151m x 17m x 5m	COGAG, 30 knots	175	2003
Destroyer – DDG	**MURASAME** (DD-101)	9	6,200 tons	151m x 17m x 5m	COGAG, 30 knots	165	1996
Destroyer – DDG	**ASAGIRI** (DD-151)	8	4,900 tons	137m x 15m x 5m	COGAG, 30 knots	220	1988
Destroyer – DDG	**HATSUYUKI** (DD-122)	2 (3)	3,800 tons	130m x 14m x 4m	COGOG, 30 knots	200	1982
Frigate – FFG	**ABUKUMA** (DE-229)	6	2,500 tons	109m x 13m x 4m	CODOG, 27 knots	120	1989
Submarines							
Submarine – SSK	**SORYU** (SS-501)	7	4,200 tons	84m x 9m x 8m	AIP, 20 knots+	65	2009
Submarine – SSK	**OYASHIO** (SS-590)	10 (1)	4,000 tons	82m x 9m x 8m	Diesel-electric, 20 knots+	70	1998
Submarine – SSK	**HARUSHIO** (SS-583)	0 (1)	3,300 tons	77m x 10m x 8m	Diesel-electric, 20 knots+	75	1990
Major Amphibious Units							
Landing Platform Dock – LPD	**OSUMI** (LST-4001)	3	14,000 tons	178m x 26m x 6m	Diesel, 22 knots	135	1998

Note: Figures in brackets refer to trials or training ships.

(114m). This provides scope for the incorporation of a vertical-launch system and a CODLAG propulsion system built around a Rolls-Royce MT-30 gas turbine. Eight frigates are eventually planned for this batch, for which further orders are planned shortly. Six additional FFX-III ships incorporating further improvements will then complete the series. Eight anti-submarine equipped AW101 Wildcat helicopters have also been ordered to operate from the frigates and four of these were delivered in June 2016. The new ships continue to replace the existing *Ulsan* and *Pohang* classes, which each saw a further vessel decommissioned on the last day of 2015.

South Korea's submarine force provides both littoral and blue-water capabilities. This is slowly being built-up to a probable total of around eighteen boats by ongoing construction of the licence-built Type 214 German design. Nine of these have been ordered in two batches shared between HHI and DSME, of which five had been delivered by mid-2016. The seventh, *Hong Beom-do*, was launched on 5 April 2016. Construction is being followed by the indigenous KSS-III type, which is likely to replace the navy's existing Type 209 submarines and will be constructed in three batches of three boats. The aim

The Republic of Korea Navy is steadily modernising its littoral warfare forces with modern vessels, notably the FFX series of frigates. Their arrival is allowing older ships, such as the 1980s/1990s-vintage *Ulsan* and *Pohang* classes, to be retired. These images show the second of the FFX-1 Incheon batch of new frigates, *Gyeonggi*, in missile-firing tests in September 2015 and the veteran *Ulsan* class vessel *Jeju* on exercises with the US Navy earlier in the year. *(US Navy)*

Table 2.2.5: REPUBLIC OF KOREA NAVY: PRINCIPAL UNITS AS AT MID 2016

TYPE	CLASS	NUMBER	TONNAGE	DIMENSIONS	PROPULSION	CREW	DATE
Principal Surface Escorts							
Destroyer – DDG	KDX-III **SEJONGDAEWANG-HAM**	3	10,000 tons	166m x 21m x 6m	COGAG, 30 knots	300	2008
Destroyer – DDG	KDX-II **CHUNGMUGONG YI SUN-SHIN**	6	5,500 tons	150m x 17m x 5m	CODOG, 30 knots	200	2003
Destroyer – DDG	KDX-I **GWANGGAETO-DAEWANG**	3	3,900 tons	135m x 14m x 4m	CODOG, 30 knots	170	1998
Frigate – FFG	FFX **INCHEON**	5	3,000 tons	114m x 14m x 4m	CODOG, 30 knots	140	2013
Frigate – FFG	**ULSAN**	7	2,300 tons	102m x 12m x 4m	CODOG, 35 knots	150	1981
Corvette – FSG	**POHANG**	18	1,200 tons	88m x 10m x 3m	CODOG, 32 knots	95	1984
Submarines							
Submarine – SSK	KSS-2 **SON WON-IL** (Type 214)	5	1,800 tons	65m x 6m x 6m	AIP, 20+ knots	30	2007
Submarine – SSK	KSS-1 **CHANG BOGO** (Type 209)	9	1,300 tons	56m x 6m x 6m	Diesel-electric, 22 knots	35	1993
Major Amphibious Units							
Amph Assault Ship – LHD	LPX **DOKDO**	1	18,900 tons	200m x 32m x 7m	Diesel, 22 knots	425	2007

is to complete each batch with progressively greater amounts of South Korean technology. The keel of the first was laid by DSME in May 2016 and is planned to enter service in 2020. Displacing around 3,700 tons submerged, her equipment includes six vertical launch tubes for cruise and – potentially – submarine-launched ballistic missiles.[13]

Turning to major surface combatants, the next significant construction effort will relate to three additional Aegis-equipped destroyers, all of which are expected to be in service by 2027. HHI has been contracted to design what is known as the KDX-III Batch II class and to build the lead ship, which will incorporate substantial improvements over the existing *Sejongdaewang-Ham* class. As for the new Japanese Aegis-equipped destroyers, some form of hybrid electric propulsion system is likely to be specified. HHI is also involved in the construction of new amphibious ships, launching the second *Cheon Wang Bong* class tank landing ship, *Cheon Ja Bong*, on 15 December 2015.

OTHER REGIONAL FLEETS

Indonesia: Indonesian naval procurement continues to be split between a small number of 'high-end' war-fighting vessels and large numbers of relatively unsophisticated patrol ships of various shapes and sizes for constabulary roles.[14] Further progress has been made in both areas over the last year. However, the long-term target of achieving a 274-strong fleet – including over a hundred combatants – by 2024 that was first announced in the Minimum Essential Forces modernisation programme initiated in 2009, looks set to be substantially scaled back to a little over 150 units. In effect, new deliveries are just about matching the pace of withdrawals.

The most costly front-line programme in progress is for the construction of three new Type 209 submarines to an improved version of South Korea's *Chang Bogo* design. The first of these, *Nagabanda*, was rolled-out in March 2016 by DSME and should be delivered on schedule during 2017. DSME is also building a second member of the class, *Trisula.* Construction then scheduled to switch to PT PAL in Surabaya where a new facility should start work on the final boat from October 2016. The selection of the Type 209 design makes sense given Indonesia already has substantial experience of operating a pair of similar boats since the early 1980s. However, it appears that other designs are also being examined for further construction that is ultimately intended

The Republic of Korea Navy is continuing to renew its amphibious forces. This picture shows the new tank landing ship *Cheon Jan Bong*, the second member of her class, shortly before being floated out by Hyundai Heavy Industries in December 2015. *(Republic of Korea Ministry of Defence)*

The Indonesian Navy's procurement is currently split between small numbers of 'high-end' vessels and larger numbers of less sophisticated patrol and missile craft. Typical of the latter is the lead KCR-60 missile-armed fast attack craft *Samapri*, seen here on exercises with the Royal Australian Navy's Wollongong in March 2016. *(Royal Australian Navy)*

Two images of the first Indonesian 'Sigma 10514' frigate *Raden Eddy Martadinata*, being floated out from PT Pal's dockyard in Surabaya in January 2016. Construction of two ships is currently underway and more could be ordered in what is probably the most important Indonesian surface warship construction programme. *(Piet Sinke via Damen Schelde Naval Shipbuilding)*

to extend the underwater flotilla to between ten and twelve submarines. These include Russia's 'Kilo' class and the French DCNS 'Scorpène' design.

Meanwhile, construction of surface combatants is being assisted by a long-standing alliance with the Dutch Damen group, which delivered four 'Sigma 9113' class corvettes between 2007 and 2009. Assembly of a follow-on class of larger 'Sigma 10514' light frigates has transferred to PT PAL, who are currently building two units. The first of these, *Martadinata*, was launched on 18 January 2016. It seems that it is currently hoped to build at least six of the class to replace the elderly *Van Speijk* class frigates, which are reported to fall due for decommissioning from 2017 onwards.

Indonesia has been happy to supplement local construction with contracts to foreign yards for specialised ships. This image shows *Rigel*, the first of two oceanographic and hydrographic research vessels built by France's OCEA to their OSV 190 SC-WB design. The aluminium-hulled vessels have a length of 60m and are operated by forty crew and scientists. *(OCEA)*

Other surface assets include the three *Bung Tomo* class corvettes originally ordered by Brunei from BAE Systems and delivered in 2014 and the three older *Fatahillah* vessels built in the Netherlands in the late 1970s. The first of these has been completing a mid-life upgrade led by Ultra Electronics but it is not known whether this work will extend to the two other ships. Fifteen Project 1331 'Parchim 1' are also still in service, although one – *Pati Ununs* – was partly sunk in a grounding incident in May 2016 and may not be repaired.

Turning to smaller vessels, recent years have seen an expansion in orders of fast attack craft based on the KCR-40, KCR-60 and KCR-63 designs. However, reports suggest that production of the KCR-63 *Klewang* class trimarans will not progress beyond a single vessel and that the SAAB combat suite envisaged for the boat will no longer be fitted. However, a fourth KCR-63 vessel has been ordered to supplement the three already in commission, whist the KCR-40 hull has been selected as the basis for a more lightly-armed patrol vessel variant.

Whilst current focus is on indigenous construction, foreign yards have been contracted for specialised ships. For example, the first of two *Rigel* class hydrographic ships built by OCEA in France arrived in Indonesia in May 2015. Her sister, *Spica* – commissioned in October of the same year – followed in December. A new sail training ship is also under construction at the Freire Shipyard in Spain for delivery in 2017.

Malaysia: A full review of recent developments in the Royal Malaysian Navy is contained in Chapter 2.2A.

New Zealand: New Zealand published a new defence white paper, a little later than planned, on 8 June 2016.[15] An update of the last white paper published in 2010, it covers a 25-year planning horizon. Whilst referencing broader tensions in Asia, the white paper focuses on New Zealand's immediate region – including Australia and the South Pacific – as well as its interests in the southern ocean. It pledges greater emphasis on maintaining awareness of an increasing busy maritime domain. Although the defence budget will remain at around one percent of GDP out to 2030, it is expected around NZ$20bn (c. US$14bn) will be available for new procurement in this timescale.

The white paper is somewhat vague about future investment. This will be given greater prominence in a new Defence Capability Plan to be set out before the end of the year. However, it appears that maritime-focused equipment will be given a degree of priority. There are commitments to renew both the small, two-ship frigate fleet and the squadron of P-3K2 maritime patrol aircraft as they fall due for replacement. Plans to procure a single support ship for diving support and hydrographic tasks are also confirmed. Most significant, however, are new pledges for a third offshore patrol vessel that will be ice-strengthened for operations in the Southern Ocean, as well as a decision to ice-strengthen the replenishment tanker that will replace the current *Endeavour*. A proposal from South Korea's HHI had previously been reported as being preferred over a rival bid from fellow Korean shipbuilder DSME for the new tanker contract. It seems that HHI's offer retains its advantage in spite of the revised specification outlined in the white paper.

North Korea: Naval news from North Korea continues to focus on the development of a potential submarine-launched ballistic missile capability. There have been at least three further underwater tests of the KN-11 missile over the past year. North Korean claims of progress are being met with considerable scepticism from South Korean and American sources. Nevertheless, the deployment of such weapons – when combined with the Democratic People's Republic nuclear programme – would constitute a material threat to the United States and its allies. It is certainly clear that the United States military is keeping a close watch on North Korea's submarine flotilla, as demonstrated by the release of reports in March 2016 that the loss of a boat operating off the North Korean coast had been noted.[16]

The K-15 missile is associated with the new indigenous 'Sinpo' class submarine, which seems to be locally designated the 'Gorae' class. It reportedly displaces in the region of 1,500 tons and incorporates one or two launch tubes in its fin. The submarine is notably larger than other indigenous boats, with North Korean strategy tending to emphasise the construction of large numbers of small submarines and fast attack craft as part of an asymmetric warfare strategy.[17] There also appears to be particular interest in exploring the use of novel hull-forms. For example, the last year has also seen some publicity being given to the relatively new 'Nongo' class of missile-armed surface effect ship. It is apparently armed with a version of the Russian 'Uran' (SS-N-25 'Switchblade') surface-to-surface missile, which has similarities with the US Navy's Harpoon. Also under construction are various iterations of high-speed, wave-piercing craft known as Very Slender Vessels (VSVs). Initially introduced to enhance infiltration capabilities, their use may be expanded to encompass other roles.

The Royal New Zealand Navy offshore patrol vessels *Otago* and *Wellington* pictured on exercises. The recent New Zealand defence white paper confirmed plans to purchase a third OPV, which will be ice-strengthened for operations in the Southern Ocean. *(Royal New Zealand Navy)*

The Philippines: The Philippine Navy's modernisation is being given impetus by developments in the South China Sea. A third *Hamilton* (WHEC-715) class high endurance cutter, *Boutwell* (WHEC-719), is being transferred from the US Coast Guard to reinforce the surface fleet and a former South Korean corvette, *Mokpo*, is also expected. Both these ships are very elderly units – if comparatively modern compared with many of the Philippine Navy's Second World War era surface combatants – but two new missile-armed frigates are also planned.

An impression of the North Korean 'Sinpo' type ballistic missile submarine, officially designated the 'Gorae' class. There has been some uncertainty as to whether she is equipped with one or two vertical launch tubes; this image shows her firing from a single launch tube in line with the most up-to-date analysis. *(H I Sutton)*

The Philippines Navy is set to take delivery of a third *Hamilton* (WHEC-715) class cutter, *Boutwell* (WHEC-719), pictured here in 2014 before her recent withdrawal from service, looks set to be transferred before the end of 2016. *(US Coast Guard)*

A version of Indian GRSE's *Kamorta* class has emerged as the preferred bidder for the contract over HHI's offer of a modified FFX-1 design on cost grounds but a deal has yet to be concluded amongst reports that there is some institutional preference for the Korean ship. In the meantime, a contract has been awarded for two AW159 Wildcat helicopters, giving the Philippine Navy an up-to-date, if modest, anti-submarine capability.

Modernisation of the amphibious force has been a key priority and clear progress is being achieved. *Tarlac*, the first of two austere amphibious transport docks based on the Indonesian *Makassar* class, was delivered by PT PAL in May 2016 and formally commissioned on the first day of the following month. A second member of the class – known as strategic sealift vessels (SSVs) in Philippine service – is under construction and a further pair may be ordered. Also commissioning on 1 June 2016 were three additional former Royal Australian Navy *Balikpapan* class heavy landing ships, which join two sister ships donated by the Australian government in 2015.

Singapore: The Republic of Singapore's rolling modernisation plan is currently seeing deliveries start of the new *Independence* class littoral mission vessels ordered from ST Engineering in early 2013. Eight ships in the class are to replace the eleven smaller *Fearless* class patrol vessels. The first arrived at Tuas Naval Base on 26 May for twelve months of further integration and testing before formal commissioning during 2017. A second, *Sovereignty*, was launched on 16 April 2016 and the current plan is to have all of the class fully operational by 2020. Core armament includes a 76mm Oto Melara gun, Mica surface-to-air missiles, an operating platform for helicopters or unmanned aerial vehicles, and lighter-calibre weapons. The design is similar to the original US Navy Littoral Combat Ship concept in being capable of operation by a small crew, with core equipment being supplemented by specific modules and associated additional personnel for specific missions. Minimal manning is also assisted by an integrated command centre combining the bridge,

The Republic of Singapore Navy is starting to take delivery of its new *Independence* class littoral mission vessels. *(Republic of Singapore Navy).*

combat information centre and machinery control room in one space. Declining birth rates means that the Republic of Singapore Navy is threatened with a manpower shortage and the highly automated approach taken by this new class appears one way to lessen the problem.

The emphasis on automation also extends to mine countermeasures (MCM). In June 2016, Singapore's Defence Minister Ng Eng Hen stated that the country was taking steps to move to a fully unmanned MCM fleet utilising autonomous surface and underwater vehicles.[18] The approach mirrors that increasingly being taken by the international MCM community given the investment in time and money, as well as the risks, associated with current, partly-manned solutions. However, few – if any – other navies aim entirely to 'take the man out of the minefield' and it will be interesting to observe whether current unmanned systems are sufficiently mature to achieve Singapore's aims.

The other major project currently underway is the acquisition of two Type 218SG submarines from Germany's TKMS. It has been confirmed that work on the first boat started in June 2015 in advance of delivery scheduled for c. 2020.

Taiwan: The last year has seen much discussion of the Republic of China Navy's fleet renewal plans, first floated towards the end of 2014. These encompass what has been termed a 'Christmas wish list' including surface escorts equipped with Aegis-like combat systems, submarines, corvettes, and amphibious shipping that could cost around US$15bn over the next twenty years.[19] The reality is likely to be somewhat less ambitious given lack of progress towards previously-stated ambitions, notably with respect to a renewed submarine flotilla.

At present, it seems that three projects are being accorded priority. Given the potential threat from mainland China, these have a largely littoral focus. The most important – and furthest advanced – is for follow-on, production variants of the *Tuo Chiang* (*Tuo River*) catamaran missile corvette delivered in December 2014. It appears that these will be built in three blocks or 'flights', each incorporating progressive improvements. The Flight I variant will reportedly feature a larger superstructure than the prototype ship and include a multi-function radar and surface-to-air missiles. However, its heavy armament of surface-to-surface missiles will be retained. Orders for the three ships in this batch are expected by 2017. Also likely to commence construction in 2017 are four high-speed minelayers, which – in similar fashion to the corvettes and existing missile-armed fast attack craft – are intended to complicate any PLAN attempt to invade across the Taiwan Strait. Also scheduled for short term construction is at least one new LPD-type amphibious transport dock to replace the existing *Anchorage* (LSD-36) class *Hsu Hai* (formerly *Pensacola*). It is notable that all three programmes are focused on domestic construction. Indeed, some commentators have suggested that the renewal plan is as much about bolstering local maritime industries as enhancing defensive capabilities.

Thailand: The Royal Thai Navy's top objective for a number of years has been to seek political approval to join other Asian fleets in acquiring a submarine force. Various attempts have foundered due to economic and practical objections but the latest attempt appears to have gained traction with the military junta currently ruling the country and is likely to receive cabinet approval.[20] If the decision is confirmed, three S26T export variants of the 'Yuan' class will be acquired over an eleven year period at a reported cost of Thai Baht 36bn (c. US$1.1bn). It appears that the submarines will operate with the Andaman Sea Fleet given the Gulf of Thailand is regarded as being too shallow for effective underwater operations.

Surface construction is dominated by the DWH300H type frigate and 'Modified River'/*Krabi* class offshore patrol vessel programmes. The former project is being carried out by South Korea's DSME under a contract for the import of a single frigate signed in 2013. First steel was cut in August 2015 and a formal keel-laying ceremony followed on 16 May 2016. Delivery remains on track for 2018. A contract has also been agreed for a second ship, which may be built in Thailand. Meanwhile, a long-expected agreement for a second offshore patrol vessel to follow *Krabi* will see BAE Systems continue to provide technical support to a local build programme at Bangkok Dock. The follow-on ship will receive a similar suite of Thales electronic equipment to the first ship but may be equipped with Harpoon surface-to-surface missiles. Further orders are planned in due course if funding allows.

Vietnam: The Vietnam People's Navy (VPN) – along with the rest of the Vietnamese armed forces – is continuing a significant build-up of its capabilities driven largely by its concerns over the possible direction of its maritime disputes with China. Russia has been its supplier of choice. Russian press reports suggest that Vietnamese contracts valued at over US$4.5bn have been received by its defence sector over recent years, encompassing submarines, surface vessels, strike fighters and advanced missile systems.

The most significant contract from a strategic perspective is undoubtedly the US$2bn deal for six Project 636.1 'Kilo' class submarines awarded in late 2009. Five of these have now been delivered to the VPN. The sixth, launched towards the end of 2015, will follow by the end of 2016. The submarines are reportedly armed with the Club-S export variant of the Kalibr (SS-N-27 'Sizzler'/SS-N-30A) cruise missile. This not only provides a significant complication to PLAN surface warships operating near Vietnam but could threaten China's network of sensitive military facilities on the island of Hainan.

The expansion of surface forces is also progressing well. The second pair of Project 1166.1E 'Gepard', being built by the inland Zelenodolsk Shipyard in Tatarstan, were launched in April and May 2016. Some reports suggest that this batch is optimised for anti-submarine operations, although it is known the design can also be equipped with the SS-N-27/30A. Previous reports have suggested that at least two

The Royal Thai Navy's 'Modified River' class offshore patrol vessel *Krabi* pictured with the US Navy cruiser *Chosin* (CG-65) in 2013. A second vessel of the class, which is similar to the Brazilian Navy's *Amazonas* and the British Royal Navy's future *Forth* class patrol vessels, has been ordered for construction at Bangkok Dock. *(Royal Australian Navy)*

more will be delivered in due course. The third major line of construction is for Project 1241.8 'Molnya' ('Tarantul V') fast attack corvettes under a contract first agreed in 2003. The deal involved a first batch of two Russian-built vessels followed by the licensed construction of a further six units in Vietnam. Both Russian ships and four of the locally-built craft have now been delivered, with the final two being launched in Ho Chi Minh City in April 2016. Talks are underway for exercise of a further batch of four, modified ships.

An interesting development in the evolving network of Asian defence ties is growing co-operation between India and Vietnam. India has been assisting Vietnam in training the crews of its new 'Kilo' class submarines. It is also looking to it as a potential purchaser of Indian defence equipment, such as BrahMos missiles and anti-submarine torpedoes. India's legacy of operating and integrating indigenous systems with Russian equipment would seem to leave it well-placed in pursuing these ambitions.

Notes

1. A good overview of the current state of the South China Sea dispute was provided by Katie Hint in an article entitled, 'Showdown in the South China Sea: How did we get here?' published on the *CNN* website on 9 March 2016 (Atlanta GA: Turner Broadcasting, 2016). A link is currently available at: edition.cnn.com/2015/10/28/asia/china-south-china-sea-disputes-explainer/

2. Details of China's activity on the Paracel Islands can be found in Victor Robert Lee's 'Satellite Images: China Manufactures Land at New Sites in the Paracel Islands', *The Diplomat* – 13 February 2016 (Tokyo: The Diplomat, 2016).

3. An excellent explanation of the significance of the OBOR strategy is contained in 'Our bulldozers, our rules', *The Economist* – 2 July 2016 (London: The Economist Group, 2016). One example of the arguments supporting an accommodation with China is set out by Pierre Marthinus in a 17 June 2015 opinion piece entitled 'See no China: Indonesia's South China Sea policy' in the *Jakarta Post* (Jakarta: PT Niskala Media Tenggara, 2016).

4. For example, see Kallol Bhattacherjee's 'Defence ties reset as Japan joins Malabar naval exercises', *The Hindu – 13 December 2015* (Chennai: The Hindu Group, 2015).

5. The quotation is from US Defense Secretary Ash Carter speaking at a Q&A session reported by Megan Eckstein in 'SECDEF Carter: China Still Invited to RIMPAC 2016 Despite South China Sea Tension', *USNI News* – 18 April 2016 (Annapolis MD: USNI News, 2016).

6. See 2016 Integrated Investment Program (Canberra: Commonwealth of Australia, 2016). A copy can be found at: http://www.defence.gov.au/whitepaper/Docs/2016-Defence-Integrated-Investment-Program.pdf

7. A summary of the principal considerations impacting the 'Shortfin Barracuda' design was provided by Gerard Autret and Sean Costello in 'Designing the Shortfin Barracuda Block 1A', *The Strategist – 8 April 2016* (Canberra: Australian Strategic Policy Institute, 2016). M. Autret is the Chief Naval Architect for the design and a DCNS expert in submarine naval architecture. The current link is: http://www.aspistrategist.org.au/designing-the-shortfin-barracuda-block-1a/

8. There was widespread reporting of the Beijing parade and the implications of the missiles displayed on the balance of power in the Pacific even if – in the editor's opinion – these somewhat over-played the vulnerabilities of US Navy carrier groups to ballistic missile attack, given considerable targeting difficulties. There was particularly comprehensive coverage in the *Financial Times*, for example Charles Clover's 'China unveils 'Guam Express' advanced anti-ship missile on 3 September 2015 (London: The Financial Times Limited, 2015). The potential future role of China's own carriers was assessed in Reuben F Johnson's 'Analysts probe role of China's future carrier force', *Jane's Defence Weekly* – 9 March 2016 (Coulsdon: IHS Jane's, 2016), p.6.

9. Web based sources the *China Air and Naval Power* blog at http://china-pla.blogspot.co.uk/ and the *China Defense Blog* at http://china-defense.blogspot.co.uk/ remain useful sources of information on the PLAN, although the former has been less active of late.

10. This development has been covered in several media sources. Ridzwan Rahmat's 'Chinese Coast Guard bolsters fleet with Type 054A missile frigate-type ship', *Jane's Navy International* – June 2016 (Coulsdon: IHS Jane's, 2016) is one good source of information.

11. There seems to be much uncertainty, even from the same publication, as to what stage deployment of China's sea-based nuclear deterrent has reached. For example, Richard D Fisher Jr.'s 'China advances sea- and land-based nuclear deterrent capabilities ' in *Jane's Defence Weekly* – 23 December 2015 (Coulsdon: IHS Jane's, 2015), p.17 quoted US Navy sources in suggesting the first patrol had already occurred. Subsequently, in the 6 June 2016 edition, Andrew Tate's 'China may be struggling to stand up sea-based nuclear deterrent' implied the first patrol had yet to occur.

12. More detail on the key items of equipment included in the Japanese 2016 defence budget can be found in the *Defense Programs and Budget of Japan – Overview of FY2016 Budget* (Tokyo: Japanese Ministry of Defence, 2016) at: http://www.mod.go.jp/e/d_budget/pdf/280330.pdf

13. Considerable detail on the KSS-III programme has been provided in news reports on the ever useful *Navy Recognition* website at: http://www.navyrecognition.com/

14. A detailed review of the Indonesian Navy's recent development was provided by Mrityunjoy Mazumdar in 'Indonesia: Lofty Ambitions tempered by Economic Realities', *Seaforth World Naval Review 2016* (Barnsley: Seaforth Publishing, 2015), pp.38–49. The *Defense Studies* blog at http://defense-studies.blogspot.co.uk/ remains a good source of Indonesian – and wider Asian – naval news.

15. See *Defence White Paper 2016* (Wellington: New Zealand Ministry of Defence, 2016). A copy can be downloaded at: http://www.defence.govt.nz/pdfs/defence-white-paper-2016.pdf

16. See Sam LaGrone's 'U.S. Official: North Korean Submarine is Missing, Presumed Sunk', *USNI News* – 11 March 2016 (Annapolis MD: USNI News, 2016).

17. More detailed analysis on the 'Sinpo' submarine can be found on the *Covert Shores* website at: www.hisutton.com/Analysis%20-%20Sinpo%20Class%20Ballistic%20Missile%20Sub.html

18. Further information is provided in Kelvin Wong's 'Singapore Navy aims for a fully unmanned mine countermeasures force', *IHS Jane's International Defence Review* – 30 June 2016 (Coulsdon: IHS Jane's, 2016).

19. A much more detailed review of Taiwan's authorised and projected naval spending plans was provided by Wendell Minnick in 'Thailand Moves on $14.7bn Indigenous Shipbuilding, Upgrade Projects', *Defense News – 23 June 2016* (Springfield VA: Sightline Media Group, 2016). These include ongoing ambitions to acquire submarines, as well as new destroyers and frigates that are all described in the report.

20. The story was covered by Wassana Nunuam in 'Navy submits B36bn plan to buy subs', *Bangkok Post* – 1 July 2016 (Bangkok: Post Publishing Public Co. Ltd., 2016).

2.2A Fleet Review

Author:
Mrityunjoy Mazumdar

THE ROYAL MALAYSIAN NAVY

A DIFFICULT BALANCING ACT AHEAD

Malaysia, with a population of around thirty-one million, has the third largest economy amongst the Association of South East Asian Nations (ASEAN). It is a nation with a peculiar territorial geography, being a country of two halves separated by some 500 miles of water – the disputed South China Sea – to which it is just one of a number of claimants. Some forty percent of its land area lies on Peninsular Malaysia (West Malaysia), with the remaining, geographically-larger eastern part comprising the regions of Sabah and Sarawak on the island of Borneo. In Borneo, Malaysia shares a long land border with Indonesia and a small border with Brunei. Thailand borders Malaysia in Peninsular Malaysia and there is also a link by causeway to the 'city state' of Singapore. With its strategic location astride the Straits of Malacca and the South China Sea, Malaysia shares maritime boundaries with almost all ASEAN countries. It has a coastline totalling some 4,675km (2,068km in peninsular Malaysia and 2,607km in East Malaysia) and claims an exclusive economic zone (EEZ) of over 590,000 square kilometres.

The Royal Malaysian Navy's F2000 type frigate *Jebat* pictured in Sydney Harbour in October 2013. The Malaysian fleet is currently primarily configured to ensure maritime security in a country whose two halves are divided by some 500 miles of water and *Jebat* is only one of four major surface combatants. However, tensions with China in the South China Sea mean that one eye has to be kept on the fleet's future warfighting potential. *(Royal Australian Navy).*

The Royal Malaysian Navy has primary responsibility for the territorial defence of these waters, acting in conjunction with the other Malaysian armed services. It also fulfils a significant constabulary role alongside the Malaysian Maritime Enforcement Agency (MMEA), the Royal Malaysian Police and other agencies in ensuring the security of the EEZ and local shipping lanes, particularly those passing through the strategic chokepoint of the Malacca Straits. In this latter role, Malaysia cooperates closely with both Indonesia and Singapore, signing the tri-national MALSINDO agreement in 2004 – later expanded to include Thailand – to co-ordinate patrol activities and share intelligence. The success of this initiative was marked by a fall in the incidence of pirate attacks from thirty-eight in 2004 to near zero levels today.[1]

Piracy is one of a number of maritime threats –

that also extend to human trafficking and illegal fishing – that are common to other countries in the region. However, the greatest and growing danger is possibly that from transnational militant groups – sometimes affiliated to the 'Islamic State' (IS) – that operate largely from the neighbouring Philippines. The 2013 Lahad Datu incident – the landing of some 235 'Sulu Sultanate' militants in Sabah – as well as kidnappings on Malaysian territory by the Philippine-based Abu Sayyaf group, have exposed serious vulnerabilities in its coastal security architecture. The former incident, particularly, left a deep impact on the national psyche, as it arguably represented only the second time Malaysia has been confronted with an external attack (the first being the 'Konfrontasi' [Confrontation] of 1963–6 that stemmed from Indonesia's opposition to the amalgamation of Sarawak with Malaysia). The creation of the joint Eastern Sabah Security Command (ESSCOM) was one significant consequence.[2] However, the rising threat from IS and its associates is a regional problem that requires co-ordinated action, a truth recognised in a joint declaration in May 2016 under which Malaysia, Indonesia and the Philippines agreed to cooperate with respect to maritime security patrols and intelligence.

So far as state-related threats are concerned, past disputes with Indonesia over the islands of Sipadan and Ligitan were settled amicably in the International Court of Justice as was the 'disagreement' with Singapore over Pedra Branca. This leaves China's increasingly assertive actions in certain Malaysian-claimed waters in the South China Sea as the main potential problem, albeit one for which there is no easy answer on Malaysia's part. Massive Chinese economic investment in Malaysia – including a recent US$2.3bn bailout of the scandal-hit 1MBD investment fund – provides China with significant financial and political leverage.[3] Nevertheless, Chinese actions are certainly influencing current defence thinking.

The RMN's early expansion was focused on smaller vessels, although the frigate *Rahmat* was finally commissioned in 1971 after a protracted building period. She is seen here leaving Portsmouth under Malaysian colours, probably on her delivery voyage. It is interesting to speculate how well the Austin Mini parked on the helicopter deck fared on the long voyage to the Far East. *(Crown Copyright 1971. Reproduced under the terms of the Open Government Licence)*

THE ROYAL MALAYSIAN NAVY'S EVOLUTION

The Royal Malaysian Navy (RMN) or *Tentara Laut Drama Malaysia* (TLDM) celebrated its eighty-second anniversary on 27 April 2016.[4] It traces its roots to the establishment of the Straits Settlement Royal Naval Volunteer Reserve (SSRNVR) in April 1934 in Singapore, then a part of British-ruled Malaya. This force was expanded in 1938 with the establishment of the SSRNVR Penang Division and was subsequently re-named the Malayan Royal Naval Volunteer Reserve in 1938. The outbreak of the Second World War the following year saw the creation of a Royal Navy Malay Section, known colloquially as the Malay Navy. Both reserve and regular personnel were to see active service during the Japanese invasion of 1941–2, with some escaping to continue the fight from what was then Ceylon. The two organisations were subsequently disbanded as an economy measure in 1947.

It was not long, however, before a naval service was re-established, doubtless spurred by the communist insurgency that became known as the Malayan Emergency. The new Malayan Naval Force (MNF) was stood up in December 1948, being headquartered at a former Royal Air Force base at Woodlands, Singapore. Development of the new force – renamed the Royal Malayan Navy in 1952 – was initially slow, even after the Federation of Malaya achieved independence from the United Kingdom in August 1957. Initial force composition was dominated by several seaward defence motor launches (SDMLs) and landing craft. The British Royal Navy continued to remain a dominant presence in the area, a situation that was to persist for the next decade and more. The Royal Malayan Navy's ships and infrastructure were formally transferred to the Federation of Malaya in July 1958. It became the Royal Malaysian Navy in September 1963 when Malaysia was created with the addition of the former British Colonies of Singapore, British North Borneo (Sabah) and Sarawak to the original federation.[5]

The country's principal maritime defence at this time continued to be provided by the British Royal Navy and its principal Commonwealth allies in the region, Australia and New Zealand, which retained a massive naval presence in the region whilst the Indonesian Confrontation was brought to a successful conclusion. During this time, the navy was headed by officers seconded from the Royal and Royal Australian Navies; it was not until the 31-year-old Indian Malaysian, Rear Admiral K Thanabalasingham, assumed office on 1 December 1967 as the navy's fifth commander that its leadership was indigenised.

By this time, the RMN was starting to benefit from significant investment. Given the Royal Navy's continued presence and the need to support the army during the period of the Confrontation, this was initially focused on coastal operations and included the transfer of second-hand mine-countermeasures vessels, as well as a series of patrol vessels built by Vosper in the United Kingdom and its local yard in Singapore. However, more potent vessels were acquired, most notably the newly-built frigate *Rahmat* in 1971. It was during this period that the foundations were laid for a much greater expansion of the RMN's capabilities following Britain's final withdrawal from Singapore in 1975.[6] The collapse of South Vietnam the same year and Vietnam's subsequent encroachment into Cambodia resulted in fears of a new external threat and a considerable expansion of the armed forces under the long-term 'PERISTA' plan announced in 1979.

The recent development of the RMN can be traced to the 1980s, when completion of new facilities at Lumut in Perak in Peninsular Malaysia coincided with the start of the navy's transition from a brown water navy to one with effective four-dimensional war-fighting capabilities and blue-water aspirations. This progressive – if not always steady – transformation has probably been best typified by the domestic construction of the MEKO 100 type *Kedah* class during the first decade of the new millennium and the induction of the RMN's first submarine, *Tunku Abdul Rahman*, in 2009. Meanwhile, a significant structural change occurred in 2005 with the creation of the MMEA coast guard organisation, to which the RMN transferred a number of patrol vessels. Most recently, in 2016, the roll-out of the '15 to 5' Fleet Transformation Programme sets a clear path for the fleet's further development.

THE ROYAL MALAYSIAN NAVY TODAY: OVERVIEW

As of April 2016, the RMN numbered some 17,000 uniformed personnel – including c. 2,400 officers – plus an additional 1,300 civilians. The ethnic composition is very predominantly ethnic Malay; there are fewer than 250 personnel from the other two main ethnic groups – Chinese and Indian – and the majority of this tiny group is in the officer corps. The RMN is actively trying to redress this situation in its ongoing recruiting efforts. In addition to the regular forces, there is also a naval volunteer reserve composed of (i) a 3,500-strong RMN Volunteer Reserve Unit – PSSTLDM – operating from eleven units around the country, (ii) a Reserve Officers Training Unit at universities and (iii) a Reserve Other Ranks Training Unit at polytechnics.

The fleet is currently approaching fifty commissioned vessels, with an average age of over thirty years. There are also seventeen Swedish-built CB-90 combat boats, growing numbers of other fast interception craft operated by the PASKAL 'Special Forces' and a handful of auxiliaries. A small naval aviation arm operates twelve helicopters. A notable problem is the myriad number of increasingly obsolescent hull types, which are becoming more difficult and increasingly expensive to keep operational, a challenge exacerbated by an inadequate operational expense budget.

As suggested by Table 2.2A.1, the navy's funding situation is mixed. Malaysia's total defence expenditure has tended to hover around 1.5 percent of GDP, increasingly markedly during the current decade on the back of favourable national economic performance. However, the budget for 2016 fell back slightly as a result of less favourable local economic expectations and higher spending priorities. The navy's total share of defence spending has inevitably fluctuated depending on the status of capital acquisitions by the three branches of the armed forces but has recently been increasing as projects such as the new littoral combat ships consume more funds. There has been a rough one-third/two-thirds split between capital and operational expenses over time. Many commentators suggest more needs to be spent if current capital and operational requirements are to be met.

The RMN fast attack craft *Ganyang* operating with the Royal Australian Navy submarine *Rankin* and the Republic of Singapore Navy *Stalwart* in May 2015. The Royal Australian Navy had a significant impact on the RMN's early development, providing three of its first four leaders, and ties remain close. There is also considerable co-operation with Singapore, particularly in ensuring security along the Straits of Malacca. Alongside the United Kingdom and New Zealand, all three countries are part of the Five Power Defence Arrangements that were established in 1971. *(Royal Australian Navy)*

ORGANISATIONAL STRUCTURE

The RMN is headquartered in Kuala Lumpur and run by an Admiralty Board headed by a four-star admiral – currently Admiral Ahmad Kamarulzaman Ahmad Badaruddin. The former Deputy Chief of Navy, he assumed office as the sixteenth Chief of Navy in November 2015 from his predecessor Admiral Abdul Aziz Jaafar. Jaafar had held office for almost eight years – the second longest tenure in the post's history. The current Deputy Chief is Vice Admiral Anuwi bin Hassan who, in turn, is assisted by several Assistant Chiefs of Staff heading up various staff branches and directorates.

The main operational naval base and home of the fleet command – named KD *Malaya* – is at Lumut, Perak. Completed in 1983, there are jetty facilities for around twenty ships and an adjacent helicopter base, KD *Rajawali*. The base is home to several

Table 2.2A.1: MALAYSIAN DEFENCE BUDGETS

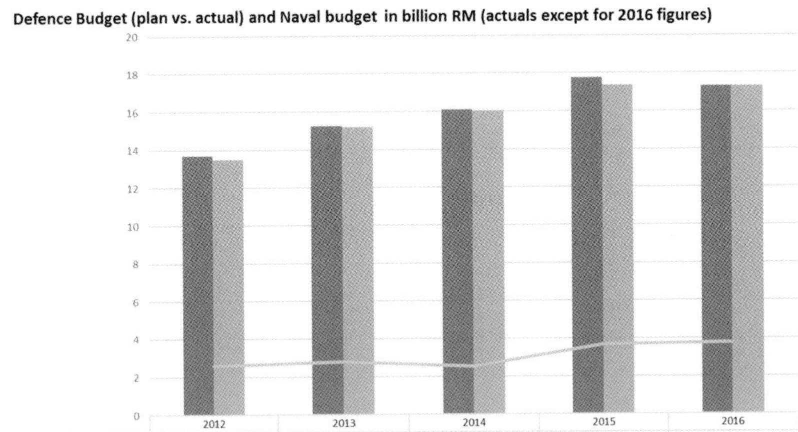

	2012	2013	2014	2015	2016
Defence Budget (plan)	13.71	15.25	16.10	17.76	17.30
Defence Budget (actual)	13.51	15.19	16.05	17.37	17.30
Naval Budget (actual)	2.58	2.78	2.52	3.66	3.77

Table 2.2A.2: MALAYSIAN NAVY BUDGETS

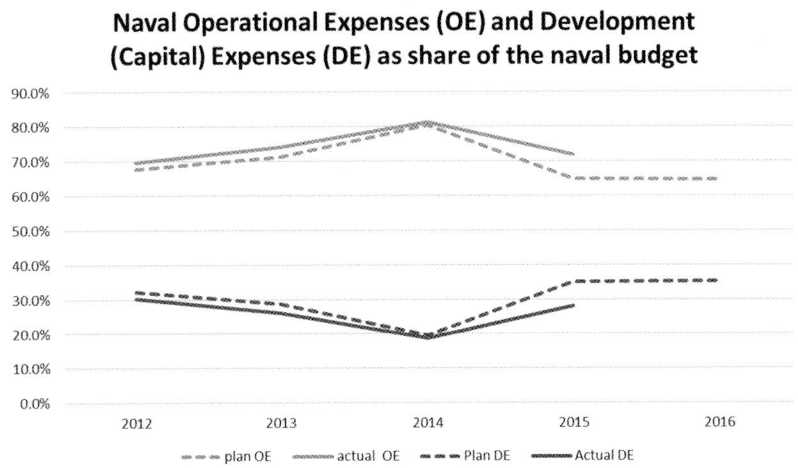

operational commands, including the training command, the naval engineering and support command and the navy's PASKAL Special Forces. The Boustead Naval Shipyard, owned by the Boustead Holdings Conglomerate, is also located next to the base and is responsible for the refit and – increasingly – the construction of many of the navy's major warships.

Geographically, there are three numbered naval regions – *Markas Wilayah Laut* (*Mawilla*) - each centred at a naval base. These are:

- Naval Region 1: Located at Tanjung Gelang, next to Kuantan Port, Pahang – a major petrochemical hub – is responsible for the western coast of Peninsular Malaysia. It became operational in 1981.
- Naval Region 2: Located at Sepanggar Bay in Kota Kinabalu, Sabah, which is home to the RMN's second largest base and, crucially, the country's submarine force. The base became operational in 2006 when the former facility at Labuan was closed. The base houses considerable support facilities for submarines and, also, a helicopter station.
- Naval Region 3: Located on Langkwai Island in Kedah, this is responsible for the northern Malacca Strait. It became operational in 2007.

There has also been discussion about the opening of a fifth major base in the oil and gas town of Bintulu in Sarawak, which was first announced in 2013. However, a report in the *Borneo Post* in May 2016 stated that the RMN was still 'contemplating' using the port as a base, suggesting it is still unclear if and when this might happen.[7]

In addition to the main regional headquarters, there are a number of forward operating bases – notably at Sandakan, Semporna and Sri Tawau in Sabah – and five Special Forces garrisons amongst the contested reefs of the Spratly Islands at the northernmost point of Malaysia's EEZ. An interesting development is the use of at least two sea bases off Sabah – a former Petronas oil platform and a

A Royal Malaysian Navy CB-90 combat boat operating with a similar vessel from the United States' Coastal Riverine 4 Squadron off the Malaysian fleet's forward operating base at Sandakan, Sabah in August 2015. The navy operates seventeen of these boats, which are often used to support Special Forces operations. *(US Navy)*

The RMN has acquired a handful of auxiliary warships to support anti-piracy missions in the Indian Ocean and maritime security operations in Eastern Malaysia. This image shows *Bunga Mas 6* in March 2015. A similar vessel, along with other assets, has been used as a sea base to support surveillance capabilities of the coast of Sabah. *(Mrityunjoy Mazumdar)*

The most significant development impacting the RMN and, indeed, potentially Malaysia's overall defensive posture in recent years has been the acquisition of two, technologically-advanced 'Scorpène' type submarines. This image shows the first of class, *Tunku Abdul Rahman*, in March 2015. *(Mrityunjoy Mazumdar)*

converted container vessel – to improve local surveillance capabilities. The hydrographic service is based at Port Klang.

All fleet operations are directed by Fleet Command Headquarters at Lumut. Afloat assets are administratively organised into several flotillas – strategic (submarines), combat, patrol, support and training (it is unclear whether the latter may have been rolled into the support flotilla). Each flotilla comprises a number of ship squadrons typically, but not always, organised by class of ship. It is believed that, uniquely, the submarine force reports directly to naval headquarters in view of the strategic nature of these platforms.

EQUIPMENT – SHIPS AND SUBMARINES

A list of principal RMN warships is provided in Table 2.2A.3. RMN ships' names are prefixed by the initials KD or *Kapal Di-Raja* meaning His Majesty's or Royal Ship.

Submarines: Long-held plans to acquire submarines finally moved forward in early 2000, no doubt influenced by neighbouring Singapore's acquisition of a similar capability. An order for two Franco-Spanish 'Scorpène' type boats and associated support was formally placed in 2002, with the first submarine being assembled by DCNS in Cherbourg and the other by Navantia in Cartagena. A 2010 Malaysian Ministry of Defence statement indicated that the two submarines cost €1.084bn (c. US$1.25bn), to which must be added €37.5m for support and test equipment and €219.3m for forty Exocet SM39 Block 2 anti-surface missiles and thirty Black Shark heavyweight torpedoes.

Prior to delivery of the two submarines, a total of 156 Malaysian personnel received training in France aboard a refurbished former French Navy *Agosta* class submarine *Ouessant*. Refurbishment was carried out between 2003 and 2005, with sea training missions taking place between November 2005 and July 2009. *Ouessant* was subsequently transported to Klebang, Malacca in Malaysia in October/November 2011 and turned into a museum exhibit.

The two submarines – *Tunku Abdul Rahman* (S630) and *Tun Razak* (S631) – were both commissioned in 2009, arriving in Malaysian waters in September 2009 and July 2010. Displacing c. 1,550 tons surfaced and a little more than 1,700 tons when dived, the two submarines have a complement of thirty-one. Published maximum speed is in the order of 20 knots (submerged) and endurance is forty-five days. There are six torpedo tubes and up to eighteen weapons can be carried. Diving depth is in excess of 350m.

Despite initial teething troubles, the boats have completed some 110 missions and exercises in seven years of service. According to the RMN, some crew members have amassed over 10,000 hours underwater. *Tunku Abdul Rahman* went in for its major repair and overhaul in November 2015 at Boustead Naval Shipyard in Lumut. *Tun Razak* is slated for a similar refit in June 2017.

The RMN was quick to invest in submarine rescue capabilities, inducting the 79m-long support vessel *Mega Bakti* in September 2013. It is operated by Kuala Lumper-based Target Mega Marine under a private finance initiative contract, using a mixed civilian and navy crew. Equipment includes a Triton XLR 09 remotely operated vehicle. From April 2016, Malaysia has also acquired a ship interface template set that allows use of the US Navy's submarine rescue equipment. The RMN submarine force has strong ties with the US Navy's submarine force and American submarines and depot ships are regular visitors to Kota Kinabalu.

Table 2.2A.3: MALAYSIAN NAVY – PRINCIPAL UNITS AS AT MID-2016

TYPE	CLASS	NO.	YEAR[1]	TONNAGE	DIMENSIONS	SPEED	PRINCIPAL ARMAMENT[2]
Submarines (2)							
Submarine – SSK	Scorpène	2	2009	1,700 tons	68m x 6m x 5m	21 knots	6 x 533mm torpedo tubes (18 torpedoes or Exocet SSM or mines)
Fleet Escorts (4)							
Frigate – FFG (L)	F2000 LEKIU	2	1999	2,400 tons	106m x 13m x 4m	28 knots	8 x Exocet SSM, 16 x Sea Wolf SAM, 1 x 57mm, 2 x 30mm, 6 x TT, 1 x hel.
Corvette – FSG	FS1500 KASTURI	2	1984	1,900 tons	97m x 11m x 4m	28 knots	8 x Exocet SSM, 1 x 57mm, 2 x 30mm, 6 x TT, 1 x hel. (platform)
A further six, 3,100-ton littoral combat ships built to DCNS' 'Gowind 2500' design have been ordered for delivery between 2019 and 2023.							
Patrol Escorts (10)							
Corvette – FSG	LAKSAMANA (ASSAD)	4	1997	750 tons	62m x 9m x 3m	36 knots	6 x Otomat SSM, 1 x quad Aspide SAM, 1 x 76mm, 2 x 40mm
Patrol Vessel – FS	MEKO 100 KEDAH	6	2006	1,850 tons	91m x 13m x 3m	22 knots	1 x 76mm, 1 x 30mm, 1 x hel. Space reserved for enhanced armament.
In addition, the elderly training frigate HANG TUAH (the former HMS MERMAID) has a secondary patrol escort role.							
Missile-Armed Fast Attack Craft (8)							
FAC – PTG	Spica M HANDALAN	4	1979	250 tons	44m x 7m x 2m	35 knots	4 x Exocet SSM, 1 x 57mm, 1 x 40mm
FAC - PTG	La Combattante II PERDANA	4	1972	250 tons	44n x 7m x 4m	37 knots	2 x Exocet SSM, 1 x 57mm, 1 x 40mm[3]
Gun-Armed Fast Attack Craft (6)							
FAC – PB	Lürssen 45 JERONG	6	1976	250 tons	45m x 7m x 3m	32 knots	1 x 57mm, 1 x 40mm
There are also an additional two slower KRIS class coastal patrol vessels, seventeen CB-90 combat boats and a number of smaller interceptors							
Mine Countermeasures Vessels (4)							
MCMV	LERICI	4	1985	600 tons	51m x 10m x 3m	16 knots	1 x 40mm
Support Vessels & Auxiliaries (6)							
LSS	SRI INDERA SAKTI	1	1980	5,000 tons	100m x 15m x 5m	17 knots	1 x 57mm, 2 x 30mm
LSS	MAHAWANGSA	1	1983	5,000 tons	103m x 15n x 5m	17 knots	2 x 57mm, 2 x 30mm
AP	SRI GAYA	2	2001	125 tons	38m x 7m x 1m	25 knots	Unarmed
AGS	MUTIARA	1	1978	1,900 tons	71m x 13m x 4m	16 knots	Unarmed
AGS	PERANTAU	1	1998	2,000 tons	68m x 13m x 4m		
There are also three converted merchant ships acting as logistic support or sea basing assets, as well as a sail training ship and three tugs. Two new training ships are under construction.							

Notes:

1 Date relates to date oldest remaining ship was first commissioned in the RMN.

2 Exocet-armed frigates and corvettes are fitted with the MM40 variant of the weapon. All the fast attack craft are believed to have the older MM38 version.

3 Two of the class have had their Exocet missiles removed and act as gun-armed fast attack craft.

Major Surface Units: There are only four relatively modern major surface combatants in the RMN. Two are the British-built 2,400-ton F2000 *Lekiu* class multi-role frigates, which were commissioned in late 1999 and currently form the 23rd Frigate Squadron. Both are slated for a mid-life upgrade that will include a new electronic support measures system – thought to be the Thales Vigile 100 Mk 2 – as well as a new electro-optronic fire-control system and Terma navigation arrays. The ship data distribution and degaussing systems will also be upgraded but there has been no news on a much-needed replacement for their increasingly obsolescent Sea Wolf surface-to-air missiles.

The *Lekiu* class are complemented by a pair of German-built 1,900-ton FS1500 *Kasturi* class corvettes that form the 22nd Corvette Squadron. They were commissioned in 1984 and have recently completed a service life extension programme. This included the installation of a new Thales TACTICOS combat management system, a Mirador electro-optical director, upgrades to the Atlas sonar system, a revised armament and torpedo launchers taken from the 'Laksamana' class corvettes.

Strike Forces: Lying somewhere between a traditional corvette and a large fast attack craft are the four heavily armed 750 ton 'Laksamana' class multi-role corvettes that constitute the 24th Corvette Squadron. They were originally ordered from Italy for the Iraqi Navy in 1981 as part of the six-strong *Assad* class but never delivered because of UN embargoes. They were acquired from builders Fincantieri under two contracts – the initial pair in July 1996 and the other two in February 1997 – after being laid up for nearly ten years. All four ships arrived in Malaysia by late September 1999. The class's combat systems are now regarded as being obsolescent but there has been some debate over the preferred supplier of upgraded equipment. Ultimately, the choice may well be down to budgetary considerations.

There are also eight more traditionally-sized fast attack craft. Four are the Swedish-built 'Spica-M' or *Handalan* class, which were all first commissioned in 1979. Armed with four MM38 Exocet surface-to-surface missiles and a 57mm gun, they have received only minor upgrades and are likely to receive new sensors soon. There are also four 'La Combattante II' or *Perdana* class vessels, which were commissioned as long ago as 1972–3 and first introduced Exocet surface-to-surface missiles into RMN service. Two – *Perdana* and *Ganyang* – serve alongside the *Handalan* class in the 1st Fast Attack Craft Squadron. The other two have shed their missiles and serve as purely gun-armed ships with the *Jerong* class (see below). Given the class is approaching its forty-fifth birthday, they are due for replacement but recent tenders for radars and directors suggest they may be retained for a while yet.

Patrol Forces: Just as is the case for the 'Laksamana' class, the six MEKO 100 type *Kedah* class corvettes that currently comprise the 17th Patrol Vessel Squadron are somewhat difficult to categorise. Relatively lightly armed in their current offshore patrol vessel role, their 1,850-ton full load displacement and MEKO modularity would allow them to be upgraded to serve as corvettes/light frigates if funded. Indeed, plans have been discussed to upgrade four of the class with improved anti-submarine capabilities, with the other two potentially receiving enhanced surface warfare weaponry.

The class originates from plans made during the 1990s to enhance existing coastal patrol vessels with much larger patrol ships. Up to twenty-seven new ships were envisaged but the Asian financial crisis at the end of the millennium and the creation of the MMEA – which assumed many of the class's intended missions – meant that only six of the 'New Generation Patrol Vessels' were actually ordered. The German Naval Group – led by Blohm & Voss – secured the contract, with the first two vessels being largely built in Hamburg before being shipped to Lumut by dock ship for final completion. The pair were both commissioned in 2006. Local construction of the four remaining ships was badly delayed by financial problems at what was then Penang Shipbuilding & Construction – Naval Dockyard but the acquisition of the business by the Boustead group allowed completion of the project during 2009–10. The ships were initially evenly-distributed across the three naval zones but four have recently been operating off Sabah due to the difficult security situation typified by the Lahad Datu incident.

The elderly 2,600-ton Type 41/61 training frigate *Hang Tuah* (the former British *Mermaid*) of the 21st Frigate Squadron retains a secondary patrol role.

Royal Malaysian Navy warships are typically organised by class of ship. *Kedah*, nearest the camera, is part of the 17th Patrol Vessel Squadron, whilst *Jebat* is part of the 23rd Frigate Squadron. They are both pictured on a live-fire exercise in August 2015. *(US Navy)*

However, the ship – which was completed in 1969 and first commissioned in 1973 – will likely be decommissioned when the first of two new-build *Gagah Samudera* vessels enter service.

Smaller patrol vessels include six elderly *Jerong* class gun-armed fast attack craft built at Hong Leong-Lürssen, Butterworth to the German joint-venture partners' designs in the 1970s. Together with the two gun-armed members of the *Perdana* class, they are believed to form the 6th Fast Attack Craft (G) Squadron. A pair of even older, *Kris*, class patrol craft – survivors of a series of twenty-four Vosper-built vessels built in three batches during the RMN's initial build-up in the 1960s – continue to operate out of Sandakan. There are also seventeen CB-90 combat boats and numbers of smaller, rigid-hulled interceptors.

Support and Auxiliary Vessels: The RMN's training assets are likely to receive a significant boost with the imminent arrival of the first of two *Gagah Samudera* class training vessels. The two ships were launched by the local NGV Tech Shipyard in December 2012 and February 2013 under a contract which saw the ships' design, hull blocks and material packages supplied by South Korea's DSME. However, completion was delayed by financial problems at the yard and final outfitting of the first ship is only now being completed at the Grade One Marine Shipyard in Lumut. Press reports suggest the ship suffered a minor engine room fire in January 2016 and this may delay commissioning. It is hoped work will resume on completing the second hull, *Teguh Samudera,* once funding is released. There is also a single sail training ship, *Tunas Samudera*.

Other significant support assets include the four *Lerici* type minehunters that comprise the 26th Mine Countermeasures Squadron. There are also two specialist hydrographic ships – *Mutiara* and *Perantau* – that were completed twenty years apart by Hong Leong-Lürssen. Amphibious capabilities were significantly diminished by the destruction of the former tank landing ship *Sri Inderapura* in a fire in October 2009. This left the two logistic support vessels *Sri Inderasakti* and *Mahawangsa* – which are also known as multi-purpose combat support ships – as the main transport vessels. They can also support smaller warships and carry out replenishment at sea, making them extremely useful assets that have been extensively deployed in support of UN missions. Two smaller fast transports built to an

An impressive overhead image of the British-built F2000-type frigate *Lekiu*, one of a pair ordered from the Yarrow shipyard on the Clyde and delivered in 1999. Plans for a second, modified pair announced in 2006 were subsequently cancelled. A 57mm gun is mounted forward, with sixteen vertical-launch cells for Seawolf surface-to-air missiles immediately aft. Four of a maximum load-out of eight MM40 Exocet surface-to-surface missiles are located between the two masts. There are 30mm guns to port and starboard of the main mast, with triple torpedo tubes on either side of the hangar. A Super Lynx helicopter is normally embarked. *(Royal Australian Navy)*

Along with the *Lekiu* class, a pair of German-built *Kasturi* class corvettes form Malaysia's force of principal surface combatants. The ships have recently completed a mid-life modernisation that has included the installation of upgraded combat systems, and an improved anti-submarine armament, as well as a revamp of gunnery armament. This 2013 image shows *Kasturi* after the modernisation was complete. *(Royal Australian Navy)*

Australian 'WaveMaster' fast ferry design serve in East Malaysia.

An interesting development in recent years has been the conversion of civilian merchant vessels to 'naval auxiliary' ships, notably the former container ships *Bunga Mas 5* and *Bunga Mas 6*. Provided and partly crewed by Malaysian International Shipping Corporation (MISC) naval reservists, both have been equipped with helicopter facilities, interceptors and military grade communications and can host PASKAL Special Forces contingents. They have largely been used as low-cost, high-endurance warships to escort and protect MISC vessels sailing in the Indian Ocean. More recently, *Bunga Mas 5* has been deployed to ESSCOM for use for sea-basing duties.[8] She joins the existing sea base *Tun Azizian* and oil platform *Tun Sharifah Rodziah* funded by the state-owned Petronas oil company through its social responsibility programme.

NAVAL AVIATION

The RMN established its first air unit – 499 Squadron – in 1988 at Lumut with six former Royal Navy Wasp HAS 1 helicopters, adding another six in 1992. This unit operated Wasps until they were withdrawn during 2001, subsequently becoming a ground support and logistics squadron.

By the late 1990s, the RMN had decided on the acquisition of Super Lynx HAS Mk.100 (also known as the Super Lynx 300) for sea control and the Eurocopter AS 555 SN Fenenc for training and utility duties. The first delivery of the Super Lynx Mk.100 took place on 3 March 2003 and were completed during 2004. Deliveries of the Fennecs also took place during 2004. Two AS 355 Ecureuil were rented between 2001 and 2004 to cover the transition phase.

The Super Lynx helicopters are operated by 501 Squadron. They are equipped with radar and a FLIR electro-optical pod but no dipping sonar. Armament includes Sea Skua missiles and light torpedoes. The Fennecs – operated by 502 Squadron – also have a maritime surveillance radar and can be equipped with a FLIR pod and door-mounted machine gun if required. Both squadrons commissioned in 2004. The primary base is KD *Rajawali* at Lumut, with Kota Kinablu forming a secondary base.

Fixed-wing maritime patrol aircraft (MPA) functions are provided by the Royal Malaysian Air Force, which operates four Beechcraft aircraft in this role.

The RMN operates a total of eight missile-armed fast attack craft of two types. This picture shows *Gempita*, one of four Swedish-built *Handalan* class boats delivered in the 1970s. Their main armament – four MM38 Exocet surface-to-surface missiles – can be clearly seen in this 2013 view. *(Royal Australian Navy)*

The multi-role support vessel *Mahawangsa*, completed in Korea in 1983, is one of two similar multi-role logistic support and transport ships that have been used extensively in support of the navy's long-distance operations. *(Mrityunjoy Mazumdar)*

SPECIAL FORCES

The RMN's 'Special Forces' Unit is known as the *Pasukan Khas Laut* (Naval Warfare Special Forces) or PASKAL. PASKAL is used for counter-terrorism, anti-piracy escort missions and protecting critical offshore infrastructure, as well as manning forward operating bases. Currently, there are two operational groups – PASKAL Unit 1 at KD *Panglima Hitam* at Lumut and Unit 2 at KD *Sri Semporna* in Sabah – along with a training unit at KD *Panglima Hitam*, also the PASKAL headquarters, Strength is classified but one recent report quoting a senior PASKAL officer suggested it is less than one percent of the navy's personnel strength, so probably no more than a few hundred strong.

Each operational unit has its own assault team, combat boat cell and combat intelligence cell. Assault teams are divided into four squadrons, which are assigned roles that are more specific and consist of a mixture of specialists. The basic tactical unit is a seven person team. Maritime assets include CB-90 combat boats and smaller interceptors, with the latter's numbers being expanded due to the problematic security situation in Sabah and the requirements of the new ESSCOM command.

PASKAL personnel wear distinctive purple berets similar to those used by Indonesia's *Kopaska* Naval Special Forces. Indeed, Indonesia was one of the country's providing training for the Malaysian unit when it was developed and the relationship between the two forces remains strong, including regular training exercises and joint deployments. In August 2015, a joint Malaysian-Indonesian Quick Response Team, operating from Tanjung Pengelih, Johor in Malaysia and at Tanjung Pinang, Indonesia was created to counter transnational crime at sea. The Malaysian element comprises two, seven person PASKAL teams and supporting helicopters.[9]

NEW CONSTRUCTION AND FUTURE PLANS

The arrival of a new Chief of Navy in the form of Admiral Badaruddin has resulted in a major re-appraisal of the RMN's future direction. The new head wants to build on the current PS1320 strategic plan (covering the period 2013-20) by undertaking a major streamlining of the navy's inventory under the '15 to 5 Transformation Plan', which is understood to have considerable political support. The basic formula is to end a negative operating cost spiral caused by the maintenance of a diverse fleet of relatively old ships by rationalising the existing force, reducing the number of major combatants from fifteen existing classes to just five major types. If realised, this would – over time – involve decommissioning all existing ships except for the new submarines and *Kedah* class new generation patrol vessels.

Under the '15 to 5' Transformation Plan, the five types of vessel forming the navy's fleet of major warships would comprise:

Littoral Combat Ships: These ships already featured in previous plans, with a variant of DCNS' 'Gowind 2500' design selected for what was originally known as the Second Generation Patrol Vessel project back in 2011. The reported programme cost is RM9bn (US$2.8bn). The keel – actually a large hull block – of the first of six ships was laid at Boustead Naval Shipyard in Lumut in March 2016. Construction of the ships is being supported by DCNS under a technology transfer and support package but equipment selection has been driven by RMN requirements. The expected completion date for the first ship is December 2019, with the other five following at six-monthly intervals through to 2023. The ships will constitute the 25th LCS Squadron.

New Generation Patrol Vessels: The *Kedah* class have proved successful in service, whilst Malaysia is understood to have the necessary intellectual property rights to permit additional construction. Continued construction would also leverage considerable amounts of investment already sunk into this project. It seems that two further squadrons of six vessels each are planned, giving a total force of eighteen ships.

Littoral Mission Ships: Little detail has been released on these ships, although it appears from RMN statements that eighteen are also planned. They would likely be required to undertake a range of patrol, mine-countermeasures and hydrographic

The RMN has a small naval aviation arm, with two squadrons operating six Super Lynx sea-control and six Fennec utility helicopters. The navy would like to acquire at least six more helicopters when funding allows. *(Mritunjoy Mazumdar)*

An image of the DCNS 'Gowind 2500' design that has been selected for the Royal Malaysian Navy's new 'littoral combat ship' light frigate requirement. The keel for the first of at least six ships was laid at the Boustead Naval Shipyard in Lumut in March 2016. Successful completion of this programme will provide a good indication of the likely performance of the broader '15 to 5 Transformation Plan'. *(DCNS)*

Selangor was the last of six MEKO 100-type New Generation Patrol Vessels to be completed, commissioning in December 2010. The Royal Malaysian Navy is considering building up to twelve more of the type under the '15 to 5 Transformation Plan'. *(Mrityunjoy Mazumdar)*

duties, suggesting a flexible mission suite, equipment and payload.

Multi-role Support Ships: These ships will replace the existing logistics support vessels and are likely to feature improved functionality for amphibious operations. South Korean and Chinese designs have been mooted to fulfil a similar requirement in the past but no firm plans have been released. Two or three ships are planned.

Submarines: Given the success of the previous programme, any expansion of the submarine flotilla would likely involve further purchases of the French 'Scorpène' type. It appears the aspiration is to operate a force of four boats.

The announcement of the '15 to 5' Plan has resulted in some uncertainty over previously planned mid-life modernisation programmes and new construction. For example, a planned contract for six 81m missile corvettes based on DSME's *Samudera Gagah* hull design may well have been overtaken by developments. The long-awaited upgrade of the 'Laksamana' class corvettes must be particularly doubtful given the aim of releasing modernisation and maintenance funds for new construction.

Fleet renewal is likely to be accompanied by modernisation of other assets. For example, at least six anti-submarine helicopters are a major priority to establish a new squadron at Kota Kinabalu. If funded, likely contenders would include the AW159 Wildcat, EC275 Super Cougar and MH-60R Sea Hawk. Other enhancements are likely to focus on a forces-wide satellite communications system to provide multi-domain awareness and, possibly, improved coastal surveillance capabilities, particularly in East Malaysia.

CONCLUSION

Although the RMN has undoubtedly made progress over the last decade, it faces a difficult balancing act ahead. The most positive development is undoubtedly the creation of the small but effective submarine flotilla, which offers deterrent capabilities far beyond its numbers. A potential adversary would have to allocate disproportionate resources to hunt just one submarine. However, limited numbers do also restrict operational flexibility, particularly when one of the boats is undergoing maintenance. Arguably, the considerable expense involved in establishing and operating a submarine force has also restricted investment in other platforms.

The fleet currently has very few modern surface combatants and faces the block obsolescence of entire ship classes, some of which are over forty years old. The '15 to 5 Transformation Plan' offers a potential solution to this problem, is regarded by the navy as 'eminently doable' and has seeming political support. In addition to inevitable questions over funding, much will depend on effective implementation, including the selection of effective platforms and combat systems. It remains an open question whether Malaysian industry can deliver the resulting ships cost-effectively and in a timely manner. The performance of the Littoral Combat Ship programme will provide clues in this regard.

A broader question relates to the Royal Malaysian Navy's appropriate focus. Given that the chances of the navy involving in conventional warfighting are relatively slim in the short to medium term, the most important duties are likely to be more limited security operations requiring numerous, economical warships. However, tensions in the South China Sea remain a complicating factor, particularly in the face of Chinese 'divide and rule' tactics. Malaysia – and the Royal Malaysian Navy – are likely to face difficult choices handling this issue in the years ahead.

The 'Laksamana' class corvette *Laksamana Hang Nadim* – a cross between a large fast attack craft and a more traditional surface escort – pictured in August 2009. Ordered by the Iraqi Hussein regime and completed in 1988, the six-strong class was never delivered to their intended owner because of UN sanctions and four were bought by Malaysia, being commissioned between 1997 and 1999. The two remaining members of the class remain laid-up at the Italian naval base at La Spezia. The Malaysian vessels are typical of the elderly, maintenance-heavy warships that dominate much of the current fleet and which the '15 to 5 Transformation Plan' aims to eradicate. Plans for modernisation may be cancelled as the RMN tries to focus investment on a limited number of modern classes. *(Royal Australian Navy)*

Notes

1. An interesting discussion of the reasons for this surprisingly effectively collaboration is provided by Terence Lee and Kevin McGahan of the National University of Singapore in 'Norm subsidiarity and institutional cooperation: explaining the straits of Malacca anti-piracy regime', *The Pacific Review* – Volume 28, Issue 4 (Abingdon: Routledge, 2015), pp. 529–52.

2. For further analysis on the impact of the Lahad Datu incident, as well as other recent influences impacting Malaysian security, please refer to Tang Siew Mun's 'New Developments in Malaysian Security Outlook' in *NIDS Joint Research Series No.11: Security Outlook of the Asia Pacific Countries and Its Implications for the Defense Sector* (Tokyo: National Institute for Defense Studies, 2014). A copy is currently available at: http://www.nids.go.jp/english/publication/joint_research/series11/index.html

3. For additional background, see Michael Peel's '1MBD: scandal-hit Malaysia fund seals China deal for power assets' in *The Financial Times* – 24 November 2015 (London: The Financial Times Ltd, 2015).

4. A more detailed overview of the RMN's historical development can be found in James Goldrick and Jack McCaffrie's *Navies of South-East Asia: A Comparative Study* (Abingdon: Routledge, 2013), pp.92–116.

5. Singapore was subsequently expelled from Malaysia two years later, forming its own armed forces.

6. The United Kingdom, Australia and New Zealand retained – and still retain – residual obligations to 'consult' with Malaysia and Singapore about threats to their security under the Five-Power Defence Arrangement of November 1971. The British Royal Navy maintains a small logistics facility at the former Singapore Naval Base at Sembawang.

7. See further Wilfred Pilo's 'Navy mulls using Bintulu Port as defence base', *The Borneo Post* – 7 May 2016 (Kuching: Borneo Post Sdn Bhd, 2016).

8. See 'Bunga Mas 5 "Mother Ship" arrives to keep Cross Border Criminals at bay' reported in the *Kinabalu Today* news portal on 22 January 2016, retrievable by searching: http://www.kinabalutoday.com/index.php

9. More detailed information on the history and status of Special Forces in Malaysia can be found in Shamsul Afkar bin Abd Rahman's thesis, *History of Special Operations Forces in Malaysia* (Monterey CA: Naval Postgraduate School, 2013) and currently available by searching Calhoun: The NPS Institutional Archive at: http://calhoun.nps.edu/

10. The RMN has produced two 'coffee table' style books – *Defining Moments*, covering the RMN's history, and *OP Fajar, The Malaysian Success Story in the Gulf of Aden* – that can be downloaded from the RMN website. Relevant links are: www.navy.mil.my/index.php/informasi/penerbitan/buku/file/135-buku-defining-moments%20a and www.navy.mil. my/index.php/informasi/penerbitan/buku/file/97-buku-op-fajar-the-malaysian-succes-story-in-the-gulf-of-aden

2.3 REGIONAL REVIEW

Author:
Conrad Waters

THE INDIAN OCEAN AND AFRICA

INTRODUCTION

Events in the Middle East continue to dominate the broader Indian Ocean and African region. Foreign intervention – most effectively by Russia in Syria – has halted the previous advances of the radicalised adherents of the so-called Islamic State. However, the conflict there, as well as those in Iraq and Yemen, progresses largely unabated. Behind much of the fighting lies a complex web of animosities – often simplified as a struggle between Shia Islam-dominated Iran and the Sunni Muslims of Saudi Arabia and its allies in the Gulf Co-operation Council and elsewhere – that defies any swift resolution. The agreement of the Joint Comprehensive Plan of Action with respect to Iran's nuclear programme on 14 July 2015 has paved the way for the end of sanctions against that country and the resumption of more normal relations with the United States and Europe. However, relations within the region remain tense. Saudi Arabia's execution of a leading Shia cleric, Sheikh Nimr al-Nimr, on 2 January 2016 saw the respected British Broadcasting Corporation (BBC) carry a report claiming that relations between Saudi Arabia and Iran were 'the most dangerous in decades'.[1]

Given this context, defence spending in the Middle East continues at a high level in spite of the impact of lower oil prices on local economies. Naval forces have typically been accorded a low priority in the area. However, there have recently been some interesting developments. Particularly remarkable has been the rapid expansion of the Egyptian Navy's capabilities. Substantial investment in new surface ships and submarines can, possibly, be explained as part of a normal process of fleet renewal. However, acquisition of the two *Mistral* class amphibious assault ships previously ordered by Russia – confirmed in September 2015 – represents a noteworthy extension in ambition for a fleet that has not been previously known for its expeditionary capabilities. Reports suggest that funding from Saudi Arabia has helped Egypt meet the €950m (US$1.1bn) purchase price. In effect, the argument goes, Saudi Arabia is looking to use the Egyptian Navy as a proxy naval force to bolster its own, rather limited maritime capabilities. This is certainly in line with the Cairo Declaration signed at the end of July 2015, under which the two countries pledged to boost economic and military ties, as well as to work together to create a joint Arab military force.[2]

It is not only Egypt that is looking at developing a more potent navy. In June 2016, Qatar announced a €4bn (US$4.5bn) deal with Italy's Fincantieri to acquire two offshore patrol vessels, four 3,000-ton 'corvettes' and its own major amphibious ship. Further contracts with Finmeccanica and MBDA for systems and missiles will reportedly increase the contract's value to nearly €6bn. The amphibious ship will be modelled on the Algerian Navy's *Kalaat Béni Abbès* and, as for Egypt's *Mistrals*, provide Qatar with a first-time expeditionary capability. How the small Qatari Navy will man these sophisticated ships is an interesting question.

Away from the Middle East, India aimed to bolster its image as the region's leading naval power with an international fleet review held off Visakhapatnam in February 2016. The first such review to take place in India since 2001, the five-day event saw ships from twenty-one foreign fleets join an Indian Navy contingent of around seventy vessels in a series of events and exercises. The review coincides with Indian Premier Narendra Modi's attempts to articulate a coherent economic and security strategy for the Indian Ocean that marks a significant shift in emphasis away from his country's previous continentalist obsession.[3] Doubtless with one eye to China's growing naval presence in the region, the strategy is based on securing the defence of India's own maritime interests whilst promoting an agenda of strengthened economic and security collaboration with its oceanic neighbours. Certainly, naval power is becoming central to India's geo-political thinking. This is also reflected in the publication of an updated naval strategy – *Ensuring Secure Seas: India's Maritime Military Strategy* – towards the end of 2015.[4] From a force structure perspective, the document is informative in confirming an intention to deploy three carrier battle groups, as well as stressing the importance of the nuclear retaliatory capability provided by the *Arihant* class strategic submarines. There is also a focus on naval procurement supporting the indigenous defence sector. Whether, in turn, industry can deliver the ships the navy requires has to be moot given a rather chequered track record.

Table 2.3.1 provides an overview of the most significant regional navies as of mid-2016.

The Algerian Navy's amphibious transport dock *Kalaat Béni Abbès* pictured on exercises with the US Navy in May 2016. Navies in North Africa and the Middle East are in the middle of a surprising expansion in amphibious capabilities. Egypt is in the course of taking delivery of two *Mistral* class amphibious assault ships previously ordered by Russia, whilst Qatar has ordered a vessel based on the Algerian ship's design. *(US Navy)*

Table 2.3.1: FLEET STRENGTHS IN THE INDIAN OCEAN, AFRICA AND THE MIDDLE EAST – LARGER NAVIES (MID 2016)

COUNTRY	ALGERIA	EGYPT	INDIA	IRAN	ISRAEL	PAKISTAN	SAUDI ARABIA	SOUTH AFRICA
Aircraft Carrier (CV)	–	–	1	–	–	–	–	–
Attack Submarine (SSN/SSGN)	–	–	1	–	–	–	–	–
Patrol Submarine (SSK/SS)	4	4	13	3	5	5	–	3
Fleet Escort (DDG/FFG)	1	7	24	–	–	10	7	4
Patrol Escort/Corvette (FFG/FSG/FS)	8	4	10	7	3	–	4	–
Missile Armed Attack Craft (PGG/PTG)	c.12	c.30	10	24	8	8	9	3
Mine Countermeasures Vessel (MCMV)	–	c.14	6	–	–	3	7	4
Major Amphibious (LPD)	1	1	1	–	–	–	–	–

Notes:

1 Algerian fast attack craft and Egyptian fast attack craft and mine-countermeasures numbers approximate.

2 Iranian fleet numbers exclude large numbers of indigenously-built midget and coastal submarines.

3 The South African attack craft and mine-countermeasures vessels serve in patrol vessel roles.

INDIAN OCEAN NAVIES

Bangladesh: The Bangladesh Navy continues to make progress with the expansion and modernisation objectives set out under the Forces Goal 2030 development plan. Support is coming from an unlikely combination of sources. The United States transferred a second *Hamilton* (WHEC-715) class high endurance cutter, *Rush* (WHEC-723), on 5 May 2015. She completed her long delivery voyage in November and was formally commissioned as *Somudra Avijan* on 19 March 2016. A third member of the class, which is currently being replaced by the US Coast Guard's new national security cutters, has been offered for transfer. The ships operate as light frigates in Bangladesh Navy service.

Also commissioned at the same ceremony as *Somudra Avijan* were the new Chinese-built corvettes *Shadhinota* and *Prottoy.* They are C13B derivatives of the domestic Type 056 'Jingdao' class and were handed-over by the Wuchang shipyard, Wuhan towards the end of December 2015. Two more ships of this class have been ordered, also to be constructed in China. There had previously been reports that they might be assembled at the domestic Khulna Shipyard, which is currently engaged in licensed production of the Chinese *Durjoy* patrol vessel design. This class is somewhat smaller – but has a greater anti-submarine warfare emphasis – than the anti-surface warfare focused C13B design. The keels of the first two ships to be fabricated in Bangladesh were laid down at a ceremony attended by Prime Minister Sheikh Hasina in September 2015. They will join two earlier units also built at Wuchang.

A key element of the Forces Goal 230 programme is the creation of a 'three dimension force' that also includes aviation assets such as maritime patrol aircraft and helicopters, as well as submarines. The former are now in service, whilst the latter capability will be provided by refurbished Type 035 'Ming' class boats, also supplied by China. In a speech marking the commissioning of *Somudra Avijan* and the two corvettes, Prime Minister Hasina confirmed that the submarines were expected to arrive before the end of 2016.[5]

Expanding alongside the navy is the Bangladesh Coast Guard, which has its own Coast Guard 2030 modernisation programme. A major step forward in terms of 'high-end' acquisitions has been the purchase of four former Italian Navy *Minerva* class corvettes. The ships are currently being refitted by Fincantieri with a lighter armament and sensor outfit reflecting their paramilitary mission. As for the submarines, delivery of all four is expected by December 2016.

India: Development of Indian maritime power has undoubtedly been a core theme of the current Modi administration's rhetoric. However, there is more room for debate as to whether this ambition has been matched by actual resources. The overall defence budget released on 29 February 2016 – whilst up only slightly year-on-year – amounted to a relatively healthy US$52bn or around 2.3 percent of GDP. Unfortunately, over twenty-five percent of this figure is allocated to pensions and an increasing proportion of the remainder to pay and benefits. This has left investment spending under pressure. Naval capital expenditure is set to decline by over

The Indian Navy's international fleet review off Visakhapatnam in February 2016 gave it the opportunity to bolster its image as a leading naval power. It was also the last hurrah of the aircraft carrier *Viraat*, first commissioned in 1959 as the British Royal Navy's *Hermes*. The carrier was withdrawn from service soon after the review and may be preserved as a floating hotel. *(Hartmut Ehlers)*

Table 2.3.2: INDIAN NAVY: PRINCIPAL UNITS AS AT MID 2016

TYPE	CLASS	NUMBER	TONNAGE	DIMENSIONS	PROPULSION	CREW	DATE
Aircraft Carriers							
Aircraft Carrier (CV)	Project 1143.4 **VIKRAMADITYA** (KIEV)	1	45,000 tons	283m x 31/60m x 10m	Steam, 30 knots	1,600	1987
Principal Surface Escorts							
Destroyer – DDG	Project 15A **KOLKATA**	2	7,400 tons	163m x 17m x 7m	COGAG, 30+knots	330	2014
Destroyer – DDG	Project 15 **DELHI**	3	6,700 tons	163m x 17m x 7m	COGAG, 32 knots	350	1997
Destroyer – DDG	Project 61 ME **RAJPUT** ('Kashin')	5	5,000 tons	147m x 16m x 5m	COGAG, 35 knots	320	1980
Frigate – FFG	Project 17 **SHIVALIK**	3	6,200 tons	143m x 17m x 5m	CODOG, 30 knots	265	2010
Frigate – FFG	Project 1135.6 **TALWAR**	6	4,000 tons	125m x 15m x 5m	COGAG, 30 knots	180	2003
Frigate – FFG	Project 16A **BRAHMAPUTRA**	3	4,000 tons	127m x 15m x 5m	Steam, 30 knots	350	2000
Frigate – FFG	Project 16 **GODAVARI**	2	3,850 tons	127m x 15m x 5m	Steam, 30 knots	315	1983
Corvette – FSG	Project 28 **KAMORTA**	2	3,400 tons	109m x 13m x 4m	Diesel, 25 knots	195	2014
Corvette – FSG	Project 25A **KORA**	4	1,400 tons	91m x 11m x 5m	Diesel, 25 knots	125	1998
Corvette – FSG	Project 25 **KHUKRI**	4	1,400 tons	91m x 11m x 5m	Diesel, 25 knots	110	1989
Submarines							
Submarine – SSN	Project 971 **CHAKRA** ('Akula II')	1	9,500+ tons	110m x 14m x 10m	Nuclear, 30+ knots	100	2012
Submarine – SSK	Project 877 EKM **SINDHUGHOSH** ('Kilo')	9	3,000 tons	73m x 10m x 7m	Diesel-electric, 17 knots	55	1986
Submarine – SSK	**SHISHUMAR** (Type 209)	4	1,900 tons	64m x 7m x 6m	Diesel-electric, 22 knots	40	1986
Major Amphibious Units							
Landing Platform Dock – LPD	**JALASHWA** (AUSTIN)	1	17,000 tons	173m x 26/30m x 7m	Steam, 21 knots	405	1971

A picture of the Indian Navy's lead Project 15A destroyer *Kolkata* (foreground) at the Indian international fleet review, with the British Royal Navy Type 45 destroyer *Defender* in the background. The two ships are both focused on the air-defence role and their designs date from roughly the same era. The British ship's two-faced Sampson radar is lighter than the four-faced MF-STAR at the Indian destroyer's foremast, allowing it to be placed higher for extended coverage, and her overall design is more stealthy. However, the Indian ship has a broader armament, with greater potential for anti-submarine and, particularly, anti-surface warfare. *(US Navy)*

ten percent to c. US$3.3bn in 2016/17. A major part of the problem appears to be a desire to expand force numbers against the flow of a trend towards smaller, more sophisticated forces that is almost universal throughout the rest of the world.[6] Difficulties with the defence procurement system would also seem to need to take their share of the blame. Evidence suggests that the Indian Ministry of Defence has struggled for several years to spend the investment budget allocated to it. Many programmes suffer years of delay even before contracts are signed. Subsequent delays are also endemic.

Whether the 'Made in India' initiative that has been given additional prominence by the current regime will be a help or a hindrance remains to be seen. There is certainly little evidence to date of a marked uptick in the patchy performance of Indian shipyards that has delayed so many naval procurement projects. In spite of this, the navy intends to extend indigenisation not only to the procurement of platforms but also to many of the items that comprise their constituent parts. The Indian Naval Indigenisation Plan 2015-30 reportedly aims to achieve local development and manufacture of forty-six key items of equipment ranging from gas turbines and surface-to-air missiles through to arrestor wires and submarine batteries.[7]

Table 2.3.2 provides an indication of current fleet strength, whilst developments with respect to core warship categories are outlined below.

Aircraft Carriers: The carrier fleet has been reduced to the solitary *Vikramaditya* following the withdrawal of the elderly *Viraat* (former Royal Navy *Hermes*) from operational service shortly after the conclusion of the international fleet review. She will be formally decommissioned before the end of 2016. The last Sea Harrier operations from her flight deck were concluded on 6 March 2016, with Indian Naval Air Squadron (INAS) 300 that operated the type subsequently transitioning to the Russian MiG-29K at a ceremony on 11 May. Meanwhile, the indigenous aircraft carrier *Vikrant* remains under construction at Cochin Shipyard. Industry appear confident that a scheduled delivery date of December 2018 will be achieved; this appears optimistic in the light of previous delays.

Major Surface Combatants: The last year has seen further progress with the Project 15A *Kolkata* destroyer programme. The lead ship carried out successful test firings of her new Indo-Israeli Barak 8 area air-defence missiles on 29 and 30 December 2015, marking a significant step forward for the navy's air defences. Sister-ship *Kochi* was commissioned on 30 September 2015 after nearly ten years under construction, still a year shorter than her elder sister. The class will be completed when a third ship, *Chennai*, is delivered in the second half of 2016. Work also continues on the follow-on Project 15B *Visakhapatnam* class. Press reports suggest the first ship should be delivered by July 2018, a little over three years after launch. As for *Vikrant*, this seems unlikely on past performance.

The other major surface programme is that for Project 17A frigates. Seven ships will be constructed by Mumbai's Mazagon Dock and Kolkata's Garden Reach Shipbuilding & Engineering (GRSE) under a contract agreed in early 2015.[8] The new ships are to be built using modular construction techniques with technical input from Italy's Fincantieri, which has also provided design assistance with the indigenous aircraft carrier. Displacing around 6,700 tons and with an overall length of 149m, the ships are only a little smaller than the Project 15A/15B series and will share the destroyer's Barak 8 air-defence system and BrahMos surface-to-surface missiles. Given delivery is unlikely to be achieved much before the mid-2020s, consideration has been given to means of maintaining total escort numbers in the meantime. Local press reports suggest that licensed construction of Project 1135.6 *Talwar* class frigates – or even purchase of the three similar Russian *Admiral Grigorovich* class ships impacted by lack of Ukrainian engines – are potential options.

Elsewhere in the surface fleet, the second Project 28 *Kamorta* class corvette *Kadmatt* was commissioned at Visakhapatnam on 7 January 2016, subsequently taking part in the international fleet review. The design of the remaining two ships has been recast to incorporate a lighter-weight composite superstructure and, potentially, other enhancements. Meanwhile, a start has been made on retiring the elderly *Veer* class attack craft, customised variants of the Soviet 'Taruntal' class. *Veer* and *Nipat* were decommissioned on 26 April 2016, leaving ten of the type in service.[9]

Submarines: Secretive trials of India's first strategic nuclear submarine, *Arihant*, continue to progress.

The Indian Navy is about to start taking delivery of its delayed 'Scorpène' type submarines, which have been built locally at Mazagon Dock in Mumbai to the French company DCNS' design under Project 75. The design also seems to be the frontrunner for the follow-on Project 75I programme. These images show the first boat in the class, *Kalvari*, before the commencement of sea trials. *(DCNS)*

Limited, probably closely-managed reports suggest, that both K-15 short range and K-4 intermediate range ballistic missiles have been successfully launched from the new submarine prior to formal commissioning before the end of 2016. Construction is also underway of improved strategic submarines under what many reports suggest will be a four- or five-boat programme. The second of these, *Aridhaman*, will possibly be ready for launch around the end of the year.

Progress with the construction of French-designed 'Scorpène' type submarines at Mazagon Dock under Project 75 has been subject to clearer reporting. The lead boat, *Kalvari*, commenced sea trials on 1 May 2016 and is due to be commissioned in the second half of the year. Five further sisters will follow by 2020. Reports suggest that – for reasons of continuity of production – a repeat order for 'Scorpène' class boats is the front-runner for the six additional submarines that are required under Project 75I. However, no final decision has been made and German, Russian and – potentially – Japanese designs are also in contention. Uninterrupted production is important to counter the block obsolescence of India's existing elderly force of Russian 'Kilo' and German Type 209 submarines. Further refits of some of the former class, which first entered service from 1986, have been authorised to extend their service lives. *Sindhukesari*, the first boat in the renewed modernisation programme, departed on a barge for Severodvinsk's Zvyozdochka yard in May 2016. Another three are likely to be refitted by the Larsen & Toubro conglomerate in India with Russian assistance; the first time such work has been carried out by a private-sector yard. Two of the German-built submarines will also be refitted at Mumbai's naval dockyard under a contract with Thyssen Krupp Marine Systems inked on 29 June 2016.

A potential fly in the ointment to the recent progress in developing India's underwater capabilities is a May 2016 decision to cancel a contract for ninety-eight Black Shark torpedoes. Supplied by Italian Finmeccanica's WASS subsidiary, these were intended to arm not only the Project 75 boats but also *Aridhaman* and the follow-on strategic submarines. The decision was taken in the aftermath of the corruption scandal related to the supply of Finmeccanica AW101 helicopters and is part of a move to blacklist further dealings with the Italian company. It leaves the submarines without a torpedo armament until a replacement can be selected and integrated, not a quick process.[10]

India's international fleet review allowed Myanmar to display its first indigenous frigate, *Aung Zeya*, to a wider audience. The ship apparently incorporates a mixture of Indian electronics and (mainly) Chinese weapons systems and has only very limited air-defence capabilities. Two follow-on ships show much greater attention to stealth design and are equipped with a helicopter hangar. *(Hartmut Ehlers)*

Other Warships: Construction of major warships continues to be supplemented by a wide range of programmes for secondary vessels. Projects making headway during 2015/16 include that for a second batch of *Car Nicobar* high speed patrol vessels. The first of these 'follow-on waterjet fast attack craft', *Tamugli*, was commissioned at Visakhapatnam on 23 May 2016. She will be joined by at least three further vessels during 2016/17. 2017 should also see the arrival of the first of five larger offshore patrol vessels being built by private-sector yard Pipavav under a contract signed in 2011. The deal marked the first major order of Indian warships from a private-sector yard. Unfortunately, Pipavav has been unable to complete the project any more effectively than its public-sector counterparts and has reportedly been fined for delays that now exceed eighteen months. The yard is now under new management and is working to complete the contract in the course of 2017.

Another notable project is that for a future minehunter class. This will encompass at least twelve vessels and is reported to be valued at as much as US$5bn. In a complex story, South Korea's Kangnam saw its original success in a 2008 tender competition set aside in 2014 for alleged illegal use of intermediaries but has now entered into a technology transfer deal with Goas Shipyard, which will build the new ships. Needless to say, these convolutions have not speeded the programme's process, which aims to deliver the class in a 2021–25 timeframe.

Longer-term plans include indigenous construction of new amphibious assault and replenishment vessels, as well as coastal anti-submarine and missile-armed fast attack craft. There has been little sign of tangible developments with any of these programmes over the last year.

Myanmar: Myanmar has been continuing its process of a largely indigenous fleet modernisation and build-up, commissioning its third locally-built light frigate, *Sin Phyu Shin*, on 24 December 2015. Like her sister, *Kyan Sit-Thar*, which was delivered in 2014, she adopts a more stealthy profile than the prototype frigate *Aung Zeya* and incorporates a helicopter hangar and improved electronics. Otherwise, principal armament and sensors appear similar to the first ship and are notable for providing a relatively weak air-defence capability. Media reports suggest that a total series of between six and eight ships is planned and it would seem subsequent vessels will incorporate further improvements. Meanwhile, India's international fleet review in early 2016 provided an opportunity to display *Aung Zeya* to a wider audience.

Simultaneous production of the smaller *Anawrahta* corvette series is also being maintained. A third ship, *Tabinshwheti*, was launched in November 2014 and she should be delivered soon. As with the frigates, she is stealthier than earlier ships and also incorporates a hangar. Other construction is focused on smaller craft, including a class of stealth fast attack craft. China and India appear to be major providers of design input and equipment, although there have been interesting reports of the development of closer defence ties with Israel. These could extend to the acquisition of Super Dvora fast patrol boats, which have also previously been supplied to India and Sri Lanka.[11]

Pakistan: The Pakistan Navy continues to discuss new naval programmes that may serve to halt the growing imbalance with India but, as previously, there have been only a few signs of tangible progress. Most significantly, reports in October 2015 suggested construction plans for Type 039A/Type 041 'Yuan' class submarines had been finalised. Construction will be carried out simultaneously in China and Karachi, with each assembling four boats, although it is not clear yet when assembly will start. Some reports suggest that the new submarines may be armed with nuclear-tipped cruise missiles, therefore giving a limited second-strike capability to act as something of a counterweight to the much greater potential in this regard provided by India's *Arihant* class.[12]

Although China remains Pakistan's defence 'partner of choice', links with Turkey are also strong. In June 2016, it was confirmed that Turkey's STM had been awarded a contract to modernise the navy's existing 'Agosta 90-B' class submarines, winning out over the boats' builder DCNS. It also seems that added impetus has been given to on/off talks to acquire four corvettes based on Turkey's 'Milgem' design, although this will be contingent on provision of lending support of up to US$400m from the Turkish government.

Sri Lanka: The Sri Lanka Navy continues a process of transition towards more 'normalised' maritime security and constabulary tasks that has been progressed since the end of its lengthy and bitter civil war in 2009. Modernisation of long-range patrol forces has been a particular priority, with India's Goa Shipyard being contracted to provide two new offshore patrol vessels in 2013. The first of these – named *Sayurala* – was launched on 10 June 2016. She is scheduled for delivery in the second half of 2017. With a full load displacement of 2,350 tons and a length of 105m, the ship is based on similar vessels already operated by the Indian Coast Guard and will ship a 76mm Oto Melara gun as its primary weapon. The arrival of the new ships will bolster longer-range patrol forces currently dominated by a handful of second-hand Indian and American units. One of these, the former Indian Coast Guard vessel *Varaha*, now *Sagara*, was formally donated in August 2015. Operated on lease since 2006, she played a decisive role alongside the other patrol vessels in disrupting Tamil Tiger maritime communications during the later stages of the civil war.

AFRICAN NAVIES

South Africa's deteriorating economic position is having a marked impact on defence resources, seemingly stalling progress with the South African Defence Review of 2014.[13] The most important naval procurement programmes are for a new survey ship (Project Hotel) and three offshore and three inshore patrol vessels (Project Biro). Tenders for these programmes closed, respectively, in June and September 2015 but no decision has yet been announced. This has apparently been because of the 'complexity' of the process. Plans to return a fourth 'Warrior' class fast attack craft to service to supplement the three existing ships used as interim patrol vessels also appear to be in abeyance. More positively, the facility at Durban was, as previously planned, returned to naval base status in December 2015. The 'Warrior' class vessel *Makhanda* is reportedly being used as a guinea pig to test the base's ability to support operations. However, considerable work is still required to realise the long-term objective of locating the new fleet of patrol ships at the facility by the 2024/5 timeframe. The underlying aim is to balance the navy's west coast-presence at Simon's Town with a facility better-placed to ensure security on the east coast following pirate activity in the Mozambique Channel. The South African-led Operation 'Copper' has done much to ensure security in this important waterway but current constraints on resources had led to the navy's presence being 'gapped' on occasion, doubtless justified by the much lower piracy threat.

Unfortunately, the reverse situation is true in the Gulf of Guinea. Here forty-four seamen were kidnapped in the first quarter of 2016 according to the International Maritime Bureau (IMB), after some improvement in the second half of 2015. As for much of sub-Saharan Africa, countries in the area have been increasing investment in constabulary patrol vessels but the process is not always going smoothly. For example, there have been reports that delivery of **Nigeria's** second Chinese-built P18N offshore patrol vessel *Unity* – a lightly-armed derivative of the Type 056 'Jingdao' class – has been considerably delayed. The original plan was to complete some outfitting at an upgraded docking facility at Port Harcourt and there have been suggestions this work is behind schedule, resulting in outfitting being transferred to China. More broadly, operational readiness in the Nigerian Navy remains poor. In an August 2015 address to navy officers, Chief of Naval Staff Vice Admiral Ibok-Ette Ibas stated that more than half of the navy's ships were out of use, blaming negligent operation for much of the problem.

In North Africa, naval procurement is dominated by Algeria and, increasingly, Egypt. Neither country appears to have a pressing need to expand their naval capabilities to the extent now underway, even if the effective collapse of centralised government in neighbouring Libya is a clear concern. The potential effect of Saudi Arabian influence in Egypt's case has already been noted, although old-fashioned 'one-upmanship' may also be playing a part.

Of the two countries, **Algeria's** modernisation programme is the more long-standing. Having commissioned the amphibious transport dock *Kalaat Béni Abbès* in March 2015, the last twelve months have seen similar ceremonies for the first of two new MEKO A-200 type frigates, *Erradii*, and the lead ship of three Chinese C28A corvettes, *Adhafer*. Both contracts apparently contain options to double class sizes. However, it is not currently known whether these will be taken up. A new mine-hunter, *El Kasseh*, was also launched by Italy's Intermarine on 5 April 2016. Future deliveries will see the arrival of two additional 'Kilo' class submarines (to make six in total) from Russia, which is also modernising a number of existing ships.

Egypt's fleet renewal started somewhat later than

Algeria's but the programme is catching up fast. In contrast to Algeria, procurement has been focused on a more limited range of suppliers. France's DCNS has largely been given responsibility for the surface fleet modernisation. In addition to the *Mistral* class ships previously referenced, the FREMM class frigate *Tahya Misr* was commissioned in June 2015 and work is underway on four 'Gowind 2500' type corvettes at both Lorient and Alexandria.[14] The first of these should be launched in September 2016. Meanwhile Germany's ThyssenKrupp Marine Systems has been given responsibility for renewal of the submarine fleet, building two pairs of new Type 2019/1400 submarines under contracts awarded in 2011 and 2014. The first boat, the prosaically-named *S41*, was launched at Kiel on 10 December 2015. An exception to this welcome homogeneity is the expected arrival in July 2016 of the Russian-built Project 1242.1 'Molnya' class fast attack craft, *R-32*. She is a unique version of the 'Tarantul' type, armed with the Moskit (SS-N-22 'Sunburn') surface-to-surface missile system. The c. 550-ton vessel was apparently gifted to Egypt by Russia in August 2015 at the time of the opening of the Suez Canal expansion project. She subsequently returned to Russia for crew-training. Most of the Egyptian Navy's existing fast attack craft are fitted with Western systems and it will therefore be interesting to see how the new ship is integrated with the rest of the fleet.

MIDDLE EASTERN NAVIES

Israel continues to maintain the most effective navy in the Middle East, largely because of the potential provided by its *Dolphin* class submarines. Expansion of the flotilla from three to six boats is well underway. The fifth, *Rahav*, was commissioned in January 2016 and *Dakar*, the final submarine, will follow by 2018. The next major programme will see delivery of four modified A-100 type corvettes from around the end of the decade. In the interim, the existing surface force is being upgraded with the roll-out of the Indo-Israeli Barak 8 area air-defence missile system and associated MF-STAR radar across

The Sri Lanka Navy offshore patrol vessels *Sayura* and *Samudura* pictured exercising with the US Navy's command ship *Blue Ridge* (LCC-19) in March 2016. The Sri Lanka Navy acquired offshore patrol vessels during the later stages of its long civil war to interdict its adversary's maritime communications and they are now a key part of the fleet as its transitions to a more constabulary-orientated role. Both Sri Lankan ships are relatively elderly – they were acquired second-hand from India and the United States – but two modern vessels are being built in India. *(US Navy)*

Egypt has embarked on a massive modernisation programme, partly assisted by Saudi Arabian funds. DCNS is supplying most of the new surface ships, which includes a sole FREMM class frigate *Tahya Misr* (originally launched as *Normandie*). This picture shows her under Egyptian colours in July 2015 shortly before her departure to Egypt. *(DCNS)*

the existing 'Sa'ar 5' and 'Sa'ar 4.5' classes. The new system was successfully trialled from the 'Sa'ar 5' corvette *Lahav* in November 2015 and is noteworthy in being able to provide an area air-defence capability from a corvette-sized platform.

As mentioned in the introduction, the deal relating to **Iran**'s nuclear programme had done little to ease regional tensions. Iran remains heavily committed to supporting ground operations by regional allies in Iraq and Syria. It is also widely regarded as having provided supplies to Houthi forces in Yemen's civil war. A number of arms seizures in the Arabian Sea by Western navies over recent months have reportedly involved Iranian weapons on the way to that country. A particular concern for the Gulf Co-operation Council forces fighting the rebels is the possibility of missile attacks on their naval forces. It appears there have already been a number of attempts in this regard using Chinese C802 surface-to-surface missiles seized from the Yemeni Navy. Iran itself continues to focus on asymmetrical capabilities such as midget submarines and missile-armed fast attack craft. The latter are supplemented by large numbers of smaller boats with rockets or simply machine guns operated by the Islamic Revolutionary Guard Corps Navy. The potential for these vessels to be used in swarm attacks has had a significant impact on the warships being acquired by other Gulf navies; for example the ability for ships to operate an embarked helicopter is of increasing importance as an effective counterweight. Anti-submarine capabilities are also more at a premium than before. This is tending to push up the size of local surface combatants compared with a previous tendency towards relatively compact vessels.[15]

Modernisation programmes amongst Gulf navies are currently at different stages. The **United Arab Emirates**, for example, has made substantial progress with fleet renewal. It is expected to complete the six-ship programme of heavily-armed *Baynunah* class corvettes by the end of 2016 and has already taken delivery of twenty-four 'Ghannatha' type multi-role vessels, as well as an Italian-built corvette and patrol ships. **Oman's** fleet recapitalisation is also well advanced, with deliveries under the 'Khareef' and 'Al-Ofouq' programmes (respectively for corvettes and offshore patrol vessels) now completed and the first of two high speed catamaran transports ordered from Austal handed over on 20 May 2016. A major joint venture has also been announced with the United Kingdom's Babcock International to develop a naval support centre at the new Duqm port that will involve a permanent British Royal Navy presence. The facility will supplement the new British base at Mina Salman Port in **Bahrain** – the first major facility opened east of the Suez Canal since 1971 – that is being constructed in a move that has dismayed campaigners critical of Bahrain's chequered human rights record.[16]

Meanwhile, modernisation amongst some of the other leading Gulf navies is less advanced. For example, the Fincantieri deal with **Qatar** will take some time to come to fruition, whilst a major recapitalisation in **Saudi Arabia** has yet to gain traction. The latter will involve the renewal of the country's Gulf-based eastern fleet based at Al Jubail in the Gulf under a Saudi Naval Expansion Program II. This looks set to involve the purchase of four upgraded *Freedom* type Littoral Combat Ships

Russia gifted its sole Project 1242.1 'Molnya' class fast attack craft to Egypt in August 2015. She subsequently returned to Russia to facilitate crew training; this image shows her being towed through the Bosphorus on 19 August 2015. It will probably not be easy to integrate her with the rest of the Egyptian fleet. *(Devrim Yaylali)*

The Royal Navy of Oman's new high speed support vessel *Al Mubshir* on trials off the Australian coast in March 2016. She was subsequently commissioned at a ceremony on 15 April 2016. She is a significantly smaller evolution of the United States Navy's *Spearhead* (EPF-1) Expeditionary Fast Transports (the former Joint High Speed Vessels). *(Austal)*

Singapore Technologies Marine has now completed interim deliveries of new offshore patrol vessels to the Royal Navy of Oman under Project Al Ofouq. This image shows the fourth and final vessel *Khassab*, during the course of sea trials in early 2016. She was subject to interim delivery in June 2016. The ships are armed with one 76mm and two 30mm Oto Melara guns and equipped with a Thales combat system and radar. There is also a helicopter platform aft. *(Singapore Technologies Marine Ltd)*

equipped with Mk 41 vertical-launch systems and six smaller corvettes, as well as additional fast patrol vessels, maritime patrol aircraft, helicopters and unmanned aerial vehicles. Negotiations with respect to this huge programme – the Littoral Combat Ship acquisition alone has been estimated at over US$11bn – are likely to be protracted and vulnerable both to changes in economic conditions and changes in defence priorities.

Notes

1. The assertion was made in a contribution by Kristian Coates Ulrichsen, an academic specialising in Middle Eastern affairs, entitled 'Iran-Saudi crisis "most dangerous in decades"', *BBC* – 4 January 2016 (London: British Broadcasting Corporation, 2016). See: http://www.bbc.co.uk/news/world-middle-east-35219693

2. Further analysis of Saudi Arabia's support for the *Mistral* class purchase was provided by Defense News in '*Mistral* Joins Egyptian Buys of French Gear', *Defense News* – 27 September 2015 (Springfield VA: Sightline Media Group, 2015). Useful insight is also gained from Bilal Khan's 'Egypt To Buy *Mistral* Helicopter Carriers', *Quwa Defence News & Analysis Group* – 27 September 2015 available at: http://quwa.org/2015/09/27/egypt-to-buy-mistral-helicopter-carriers/

3. The significance of the focus of Prime Minister Modi's administration on maritime power is explained in C. Raja Mohan's 'Narendra Modi and the Ocean: Maritime power and responsibility', *The Indian Express* – 12 March 2015 (Mumbai: Indian Express Group, 2015). The international fleet review was attended by over fifty nations when delegations – as opposed to just countries sending ships – are included. A good overview of the event's importance is contained in Rupakjyoti Borah's 'The Indian Navy Charts A New Course With Its 2016 International Fleet Review', *The Diplomat* – 16 February 2016 (Tokyo: The Diplomat, 2016).

4. See further, *Ensuring Secure Seas: India's Maritime Military Strategy – Naval Strategic Publication (NSP) 1.2* (New Delhi: Indian Navy, 2015). A copy can currently be downloaded at: http://indiannavy.nic.in/sites/default/files/Indian_Maritime_Security_Strategy_Document_25Jan16.pdf

5. Prime Minister Hasina was quoted in an unattributed article dated 19 March 2016 entitled, 'Navy to get 2 submarines this year: PM' in the *Daily Sun* (Dhaka: East West Media Group, 2016).

6. For further analysis of the defence budget, see Laxman K Behera's, *All About Pay and Perks: India's Defence Budget 2016-17* (New Delhi: Institute for Defence Studies and Analyses, 2016). He quotes Premier Modi as stating, at the Combined Commanders Conference in December 2015, that 'We need forces that are agile, mobile and driven by technology, not just human valour'.

7. See further Rahul Bedi's 'Indian Navy announces 15-year indigenisation plan', *Jane's Defence Weekly* – 29 July 2015 (Coulsdon: IHS Jane's, 2015), p.8.

8. Four will be built by Mazagon Dock and three by GRSE.

9. There were originally thirteen members of the class. One, *Prahar*, was sunk after a collision in April 2006.

10. For criticism of the decision's implications, see Sandeep Unnithan's, 'Contract for 98 Black Shark torpedoes scrapped, Indian nuclear submarine programme further delayed, *Mail Today* – 1 June 2016 (New Delhi: India Today Group, 2016). The editor is grateful to Mrityunjoy Mazumdar for bringing this article to his attention.

11. The Myanmar Navy was covered in some depth by Mrityunjoy Mazumdar in 'Bangladesh And Myanmar – Resurgent Navies in the Bay of Bengal: Friendly Rivalry or Something More?', *Seaforth World Naval Review 2015* (Barnsley: Seaforth Publishing, 2015) pp.50–9. Another, slightly more recent piece of research is provided by Stefan Tan Ying Xian in 'The future of Myanmar's Naval Modernization: Analysis', *Eurasia Review* – 25 March 2016 (Albany OR: Buzz Future LLC, 2016).

12. See Usman Ansari's 'Pakistan, China Finalize 8-Sub Construction Plan', *Defense News* – 11 October 2015 (Springfield VA: Sightline Media Group, 2015).

13. Much information on ongoing developments with respect to the South African and other African navies can be found on the naval portal of the *defenceWeb* news site, located at: http://www.defenceweb.co.za/

14. The lead Egyptian *Mistral*, named *Gamal Abdel Nasser* (the former *Vladivostok*), was handed over on 2 June 2016. She arrived in Alexandria on the 23rd of the same month. She will be followed by *Anwar El Sadat* (the former *Sevastopol*) in the second half of 2016.

15. A much more detailed analysis of current procurement by the Gulf Co-operation Council States than there is space to provide in this review is contained in Kate Tringham's and Alex Pape's 'Expansion and renewal: Naval procurement in the Gulf', *Jane's Defence Weekly* – 2 March 2016 (Coulsdon: IHS Jane's, 2016), pp.28–33.

16. The ongoing human rights concerns relating to this decision were highlighted by Jamie Merrill in 'Royal Navy base construction begins in Bahrain as Britain seeks a return to 'East of Suez'', *The Independent* – 1 November 2015 (London: Independent Digital News & Media, 2015).

2.4 REGIONAL REVIEW

EUROPE AND RUSSIA

Author:
Conrad Waters

INTRODUCTION

Just as China's assertive stance to its maritime claims is dominating the course of naval developments in Asian waters, so Russia's willingness to deploy force in support of its political objectives is having a major impact on navies throughout Europe. 2015 saw the war in Ukraine's Donbass region – in which Russia is widely regarded as having played a decisive role in support of rebel forces – settle down into an ongoing but low-level 'frozen conflict'. However, this de-escalation was more than counterbalanced by the start of Russia's high-profile intervention in Syria's Civil War from September 2015. This decisively reversed a series of setbacks previously experienced by Syrian government forces. The intervention saw the first operational use of Russia's long-range Kalibr-NK cruise missiles. These were fired from surface vessels in the Caspian Sea and from the newly-commissioned submarine *Rostov-on-Don* in the Mediterranean against opposition targets. Although probably of limited military value given the heavy Russian Air Force presence already deployed in Syria, the attacks did powerfully demonstrate that the Russian Navy is able to wield an equivalent capability to the American Tomahawk should the circumstances require. Indeed, this demonstration was probably the engagement's main aim.

Russia's Syrian adventure comes at a time when its relations with NATO remain tense. From a Russian perspective, NATO 'expansionism' during the post-Cold War period has reduced its own influence and continues to threaten its security. Current Russian policy is therefore influenced by the need to reverse this trend, an imperative readily seen in the new Russian 'Maritime Doctrine' announced by President Vladimir Putin during the annual Navy Day celebrations on 26 July 2015.[1] The new doctrine revises a previous version dating from 2001. It covers four functional areas – naval activity, marine transport, maritime science and resources development – as well as six regions, viz. the Atlantic (including the Mediterranean), Arctic, Pacific, Caspian Sea, Indian Ocean and Antarctica. Of these regions, priority is being given to the Atlantic and Arctic. The attention on the Atlantic – which includes a desire to integrate the re-acquired Crimea into the Russian economy and re-establish a permanent presence in the Mediterranean – is explicitly justified by NATO enlargement. The focus on the Arctic – probably equally unwelcome to the West – is explained by the region's vast mineral resources and the growing importance of the Northern Sea Route as ice caps shrink. Another key priority is the advancement of the shipbuilding section, in which substantial investment is being made. However, claims '… naval shipbuilding is doing work on a scale comparable to what was happening in the Soviet period' smack of hyperbole given recent setbacks suffered as a result of disruption of components previously supplied from Ukraine.[2]

The perspective of many of Russia's European neighbours on recent events is often different. Russia's various military deployments – as well as incidences of harassment of NATO forces in the Black Sea and Baltic – are seen as marking the return of a threat that many hoped had been permanently eradicated by the Cold War's end. As previously highlighted, defence budgets are rising in response. However, resources remain stretched. Hard choices have often to be made between funding the conventional forces required to counter potential Russian aggression and the security apparatus required to combat the actual attacks perpetrated by Islamic extremists in Paris, Brussels and elsewhere.

A recent development that reflects both these main challenges faced by the NATO alliance was the opening of the initial European Aegis Ashore facility at Naval Support Facility (NSF) Deveselu in Romania on 12 May 2016. A shore installation of Commander, Navy Installations Command and manned by US Navy sailors, NSF Deveselu is built around a SPY-1D multi-function radar installation and three eight-cell Mk 41 launch systems. Part of the European Phased Adaptive Approach (EPAA) to ballistic missile defence announced by US President Barack Obama in 2009, the facility supplements Aegis-equipped destroyers already deployed to Europe. It will be joined by a similar site in Poland in 2018. EPAA was ostensibly put in place to defend Europe from the risk of extremist attack resulting from the proliferation of ballistic missiles in the hands of countries like Iran. However, the approach has been heavily criticised by Russia for potentially reducing the effectiveness of its own strategic missile systems that underlie the concept of mutually-ensured destruction.[3]

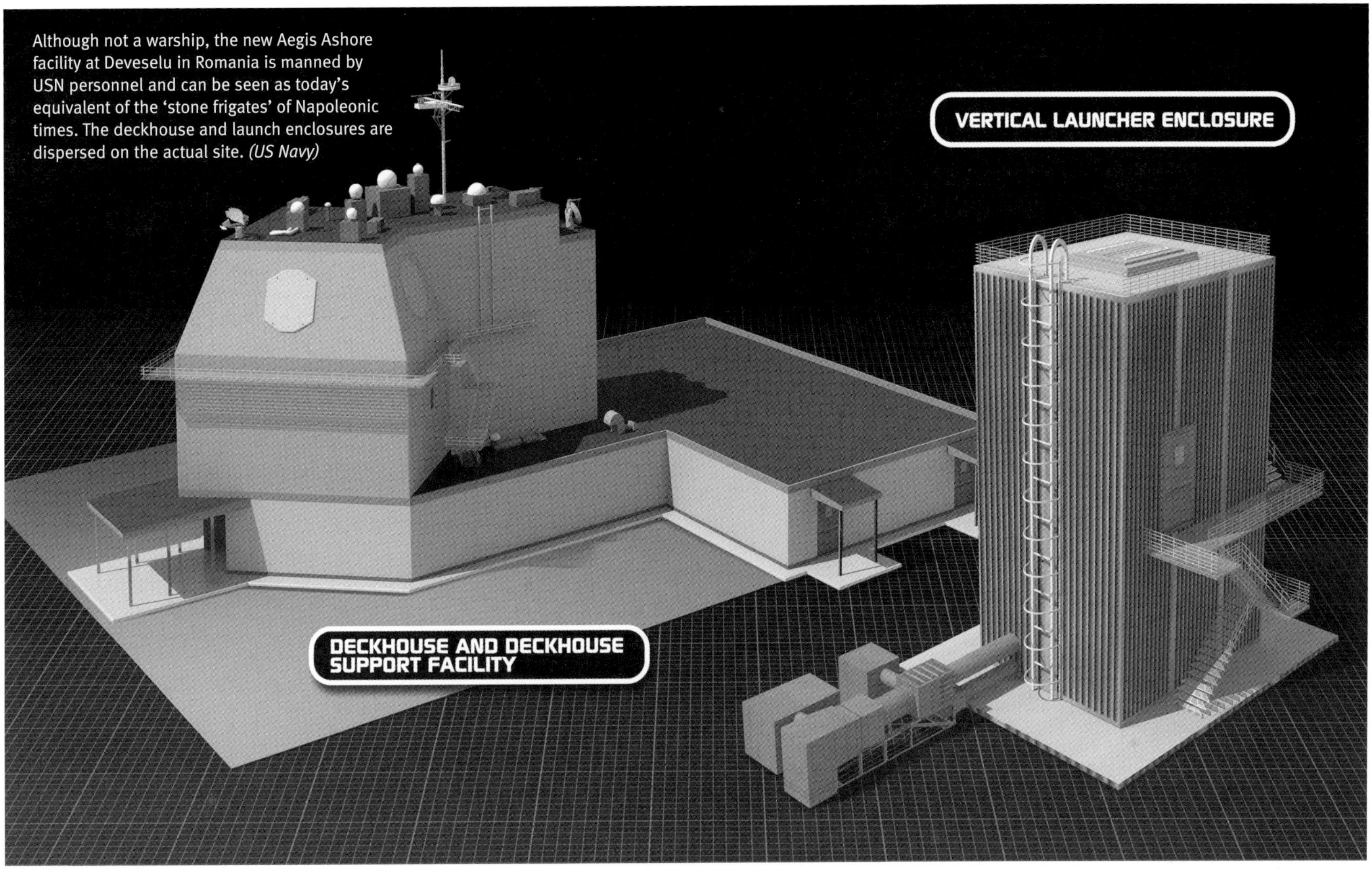

Although not a warship, the new Aegis Ashore facility at Deveselu in Romania is manned by USN personnel and can be seen as today's equivalent of the 'stone frigates' of Napoleonic times. The deckhouse and launch enclosures are dispersed on the actual site. *(US Navy)*

TABLE 2.4.1: FLEET STRENGTHS IN WESTERN EUROPE – LARGER NAVIES (MID 2016)

COUNTRY	FRANCE	GERMANY	GREECE	ITALY	NETHERLANDS	SPAIN	TURKEY	UK
Aircraft Carrier (CVN/CV)	1	–	–	1	–	–	–	–
Support/Helicopter Carrier (CVS/CVH)	–	–	–	1	–	–	–	–
Strategic Missile Submarine (SSBN)	4	–	–	–	–	–	–	4
Attack Submarine (SSN)	6	–	–	–	–	–	–	7
Patrol Submarine (SSK)	–	5	11[2]	6	4	3	13	–
Fleet Escort (DDG/FFG)	17	10	13	18[2]	6	11	16	19
Patrol Escort/Corvette (FFG/FSG/FS)	15	5	–	4	–	–	8	–
Missile Armed Attack Craft (PGG/PTG)	–	4	17	–	–	–	20	–
Mine Countermeasures Vessel (MCMV)	14	10[1]	4	10	6	6	15	15
Major Amphibious (LHD/LPD/LPH/LSD)	3	–	–	3	2	3	–	6

Notes:

1 Two further units used as support vessels.

2 Headline figures overstate the actual position, as old ships will be withdrawn once new units are fully operational.

The first 'Barracuda' type nuclear-powered attack submarine, *Suffren*, pictured under construction at DCNS' Cherbourg facility in April 2016. Her launch is scheduled for 2017. *(DCNS)*

D'Entrecasteaux is the first of four B2M multi-role support vessels being built by the Kership alliance of Piriou and DCNS for use in France's overseas departments. Although lightly armed, she is able to transport troops and supplies over reasonable distances and disembark them when there are only austere harbour facilities available. *(DCNS)*

MAJOR REGIONAL POWERS – FRANCE

The last twelve months have seen the *Marine Nationale* progress existing equipment programmes without making significant new announcements. Operationally, the carrier *Charles de Gaulle* has been active conducting strikes against Islamic State terrorists, commanding the US Navy's Task Force 50 for a time. Her November 2015 deployment to the Middle East in response to the terrorist attacks in Paris the same month saw the last operational deployment of the venerable Super Étendard strike fighter. Final launches from the carrier were made in May 2016 and the aircraft is to be formally withdrawn from service in July.

Table 2.4.2 shows there has been little change to fleet composition year-on-year. Major developments with principal ship types are highlighted below.

Aircraft Carriers and Amphibious Ships: Whilst *Charles de Gaulle* has seen considerable service over the past year, she is scheduled to undergo her second major refit and refuelling at DCNS Toulon during 2017–18. The eighteen-month package of work will encompass upgrades to electronic systems and modifications to optimise the ship for an all-Rafale strike group as well as overhaul of existing plant and machinery. The programme – which should allow a further ten years of service before further refuelling is necessary – will leave France without an operational aircraft carrier. There have been reports that at least one of the navy's Rafale squadrons may deploy on a US Navy *Nimitz* class vessel while the refit takes place to maintain their proficiency.

The dock landing ship *Siroco* decommissioned as planned in July 2015 and was subsequently sold to Brazil. The three remaining BATRAL type landing ships are also likely to be withdrawn soon, with decommissioning of *La Grandière* scheduled for July 2016. Their replacements – the B2M *bâtiment multi-mission* type of the *D'Entrecasteaux* class – are described in more detail below.

Surface Combatants: Surface construction is currently dominated by ongoing deliveries of the FREMM class multi-mission frigates. The third ship, *Languedoc*, was delivered on 16 March 2016 and the fourth, *Auvergne*, launched on 2 September 2015. Construction of the remaining four French ships in the class will continue through to 2022, when assembly of the new lighter and cheaper

TABLE 2.4.2: FRENCH NAVY: PRINCIPAL UNITS AS AT MID 2016

TYPE	CLASS	NUMBER	TONNAGE	DIMENSIONS	PROPULSION	CREW	DATE
Aircraft Carriers							
Aircraft Carrier – CVN	CHARLES DE GAULLE	1	42,000 tons	262m x 33/64m x 9m	Nuclear, 27 knots	1,950	2001
Principal Surface Escorts							
Frigate – FFG	AQUITAINE (FREMM)	3	6,000 tons	142m x 20m x 5m	CODLOG, 27 knots	110	2012
Frigate – FFG	FORBIN ('Horizon')	2	7,100 tons	153m x 20m x 5m	CODOG, 29+ knots	195	2008
Frigate – FFG	CASSARD (FAA-70)	2	5,000 tons	139m x 15m x 5m	CODAD, 30 knots	250	1988
Frigate – FFG	GEORGES LEYGUES (FASM-70)	5	4,800 tons	139m x 15m x 5m	CODOG, 30 knots	240	1979
Frigate – FFG	LA FAYETTE	5	3,600 tons	125m x 15m x 5m	CODAD, 25 knots	150	1996
Frigate – FSG	FLORÉAL	6	3,000 tons	94m x 14m x 4m	CODAD, 20 knots	90	1992
Frigate – FS[1]	D'ESTIENNE D'ORVES (A-69)	9	1,300 tons	80m x 10m x 3m	Diesel, 24 knots	90	1976
Submarines							
Submarine – SSBN	LE TRIOMPHANT	4	14,400 tons	138m x 13m x 11m	Nuclear, 25 knots	110	1997
Submarine – SSN	RUBIS	6	2,700 tons	74m x 8m x 6m	Nuclear, 25+ knots	70	1983
Major Amphibious Units							
Amph Assault Ship – LHD	MISTRAL	3	21,500 tons	199m x 32m x 6m	Diesel-electric, 19 knots	160	2006

Note:
1 Now officially reclassified as offshore patrol vessels.

frégate de taille intermédiaire is expected to be well underway. Five of the class are expected to be delivered from 2023 onwards to maintain the currently planned surface force of fifteen major combatants. Few details of the new ships are currently available, with DCNS likely to reveal the design of the new ships at the Euronaval exhibition in the autumn of 2016.

Submarines: The pattern of future French submarine construction is also well-established. Construction throughout the next decade and beyond is dominated by ongoing assembly of the 'Barracuda' type nuclear-powered attack submarines of the *Suffren* class. Construction of the lead boat is progressing a little more slowly than first planned and she will now be launched in the second half of 2017. Meanwhile, work is drawing towards a close on modernisation of the *Le Triomphant* class strategic submarines to accommodate the new M51 ballistic missile. The fourth boat, *Le Terrible*, was designed to house the missile from the build but the other three submarines have been going through a modification programme at Brest. The newly-modified *Le Triomphant*, the second boat to go through modernisation, is due to undertake a test launch of the missile in the second half of 2016. The programme will conclude when *Le Teméraire* returns to the fleet in 2018.

Minor Warships and Auxiliaries: In the absence of a comprehensive programme of new offshore patrol vessel construction, replacement of the second-line naval vessels that are largely based in France's overseas departments is being implemented by the piecemeal acquisition of support-type vessels with marginal combat potential. The Kership alliance between Piriou and DCNS is playing a leading role in these programmes. The first *D'Entrecasteaux* class B2M logistic support vessel was delivered from the Concarneau yard in March 2016 and will be based in Nouméa in New Caledonia. Three follow-on vessels will be deployed to Papeete in Tahiti, Port-des-Galets in La Réunion and Fort-de-France in Martinique.[4] The relatively large 2,300-ton design is armed only with light machine guns but is equipped with good communication facilities and can deploy a small military force of up to twenty troops and supporting light vehicles. A 12-ton capacity crane (at 14m extension) provides additional flexibility when operating from austere harbour facilities. Kership has also received orders for four BSAH *bâtiments de soutien et d'assistance hauturiers* that have some similarities with the B2M design.

MAJOR REGIONAL POWERS – ITALY

The future direction of Italy's *Marina Militare* will be heavily influenced by a new white paper on defence published in April 2015. Arguably the first full review of Italy's defence policy since 1985, it places emphasis on giving the Italian armed forces a central role – backed up by full spectrum military capabilities – in the broader Mediterranean area (the 'Euro Mediterranean' region). Operations beyond this region will, in contrast, be restricted to contributing to missions led by allied powers.[5] The new strategy appears sensible given the ongoing humanitarian refugee crisis in the Mediterranean and the instability present along much of its eastern and southern shores. For the navy, the longer-term consequence will likely be some re-orientation towards regional intervention capabilities at the expense of its current enthusiasm for extra-regional expeditionary deployments.

Medium-term naval prospects continue to be dominated by the exceptional €5.4bn (c. US$6bn) funding package approved under a 'Naval Law' at the end of 2014 that will permit renewal of a significant part of the surface fleet. The last twelve months have seen significant progress achieved in finalising the designs and construction of the various ships in the programme. These include a LHD-type amphibious assault ship, a new fleet replenishment vessel, a series of frigate-like multi-role patrol vessels and two high-speed Special Forces insertion craft. It has become apparent that many of the new ships will be larger and more potent than was initially apparent. This has resulted in criticism from parts of

The Naval Law authorised at the end of 2014 is leading to a surge of Italian warship procurement. In addition to a series of PPA type frigates, this has included orders for a new amphibious assault ship and a logistic support replenishment vessel. The specifications of the amphibious ship, to be named *Trieste*, have steadily expanded and the current twin-island design has similarities with the British Royal Navy's *Queen Elizabeth* class carriers. The navy's emphasis of the vessel's humanitarian potential has resulted in some incredulity from the Italian press. *(Italian Navy via Rolls-Royce / OCCAR)*

the press that Italian politicians were duped into approving a fleet of powerful warships that were 'sold' to them as largely constabulary assets with a strong humanitarian focus.[6]

The most impressive ship in the new package is the new amphibious assault ship, which is to be named *Trieste*. Ordered from a consortium comprising Fincantieri and Finmeccanica at a cost of c. €1.1bn (US$1.2bn) in July 2015, the ship is likely to be built at the Castellammare di Stabia yard near Naples for delivery in 2022. Although details remain a little sketchy, recent reports suggest that *Trieste* will become the largest ship in the Italian fleet with a length of some 245m and a full-load displacement of around 32,000 tons. Different conceptual images have portrayed single and double-island designs but it appears that it is the latter arrangement, which has similarities with the British *Queen Elizabeth* class aircraft carriers, that has found favour. The ship's combined diesel or gas (CODOG) propulsion arrangement will also include two of the Rolls-Royce MT-30 gas turbines that are also used in the British vessels. Although apparently lacking a ski-jump, the LHD-design has the space and fuel capacity to undertake a secondary aircraft carrier role; indeed some commentators expect a ski-jump to be added to the design at a later date.

The award of the *Trieste* propulsion contract to Rolls-Royce was something of a consolation prize for the loss of its bid to power the new PPA (*Pattugliatore Polivalente d'Altura*) patrol vessels. These will be equipped with an innovation hybrid propulsion plant comprising a GE LM2500+ gas turbine for speeds of up to 33 knots, two diesels for mid-range operation and two gearbox mounted electric motors for low speeds. An order for six of these ships and four options was placed in May 2015; one of the options was subsequently exercised in November of the same year. The class is designed to be completed in various configurations ranging from 'Light' (constabulary) to 'Full' (war-fighting) utilising a common hull. All versions include 127mm and 76mm guns, with the higher-specified ships incorporating a more comprehensive radar suite and vertical launch cells for Aster missiles. The orders exercised to date include two ships in 'Full' specification, three in 'Light-Plus' configuration and two 'Light' variants. Costs – including a ten year support contract – range from around €530m (US$600m) for the 'high end' vessel down to €430m (US$475m) for the 'Light' specification.[7]

The other vessels authorised under the Naval Law have also been ordered. The new 165m-long replenishment vessel is being fabricated in two sections by Fincantieri's facilities at Castellammare di Stabia and Riva Trigoso, where work started in February 2016. The two sections will be transported for integration at the Muggiano yard near La Spezia in a process similar to that previously used for the aircraft carrier *Cavour*. Delivery is scheduled for 2019. The contract for the two high-speed insertion craft, also known as multi-role ships, has been allocated to neighbouring Intermarine under a €40m (US$44m) deal confirmed in March 2016.

Given that delivery of the majority of ships approved under the Naval Law is some years away, the navy's immediate structure is largely dependent on the delivery of FREMM frigates and Type 212A submarines approved under previous programmes. The former are being constructed at the rate of approximately one ship p.a. The sixth vessel, *Luigi Rizzo*, was the latest to start sea trials when she departed Fincantieri's Muggiano yard in May 2016. Meanwhile, delivery of the third Type 212A submarine, *Pietro Venuti*, is expected to take place during July 2016. The last boat in the series, *Romeo Romei*, was launched on 4 July 2015 and she should commission in the next twelve months. As indicated by Table 2.4.3, however, the new arrivals are being increasingly outweighed by withdrawals of older ships. These have included the decommissioning of

Table 2.4.3: ITALIAN NAVY: PRINCIPAL UNITS AS AT MID 2016

TYPE	CLASS	NUMBER	TONNAGE	DIMENSIONS	PROPULSION	CREW	DATE
Aircraft Carriers							
Aircraft Carrier – CV	CAVOUR	1	27,100 tons	244m x 30/39m x 9m	COGAG, 29 knots	800	2008
Aircraft Carrier – CVS	GIUSEPPE GARIBALDI[1]	1	13,900 tons	180m x 23/31m x 7m	COGAG, 30 knots	825	1985
Principal Surface Escorts							
Frigate – FFG	CARLO BERGAMINI (FREMM)[2]	5	6,700 tons	144m x 20m x 5m	CODLOG, 27 knots	145	2013
Frigate – FFG	ANDREA DORIA ('Horizon')	2	7,100 tons	153m x 20m x 5m	CODOG, 29+ knots	190	2007
Destroyer – DDG	DE LA PENNE	2	5,400 tons	148m x 16m x 5m	CODOG, 31 knots	375	1993
Frigate – FFG	MAESTRALE	7	3,100 tons	123m x 13m x 4m	CODOG, 30+ knots	225	1982
Frigate – FFG	ARTIGLIERE	2	2,500 tons	114m x 12m x 4m	CODOG, 35 knots	185	1994
Frigate – FS	MINERVA	4	1,300 tons	87m x 11m x 3m	Diesel, 25 knots	120	1987
Submarines							
Submarine – SSK	TODARO (Type 212A)	2	1,800 tons	56m x 7m x 6m	AIP, 20+ knots	30	2006
Submarine – SSK	PELOSI	4	1,700 tons	64m x 7m x 6m	Diesel-electric, 20 knots	50	1988
Major Amphibious Units							
Landing Platform Dock – LPD	SAN GIORGIO	3	8,000 tons	133m x 21m x 5m	Diesel, 20 knots	165	1987

Note:
1 Now operates largely as a LPH.
2 Class includes BERGAMINI (GP) and FASAN (ASW) variants.

The Italian *Maestrale* class frigate *Aliseo* pictured refuelling from the German oiler *Spessart* during a NATO deployment in the Adriatic in 2015. The early 1980s-era *Maestrale* class are now starting to be replaced by the new FREMM class. *(US Navy)*

the lead *Maestrale* class frigate, the 'Soldati' class frigate *Grantiere* and a further two *Minerva* class corvettes over the past year.

The *Enrica Lexie* case – referenced in several previous editions of *Seaforth World Naval Review* – involving the detention of two Italian marines for killing two Indian fishermen mistaken for pirates when they approached the tanker in February 2012 – has been referred to international arbitration. The two personnel involved – one of whom suffered an ischaemic attack in 2014 – have now both been returned to Italy pending resolution of the arbitration proceedings.

MAJOR REGIONAL POWERS – SPAIN

The *Armada Española* has essentially been marking time over the last few years given that the impact of the Eurozone crisis on defence spending halted all but the most essential defence procurement. Fortunately, there has been a steady return to greater financial stability and this has allowed some long-delayed naval procurement programmes to be revisited.

The main future surface programme is that for the new class F-110 frigates. Five of these are to be delivered between 2022 and 2030 to replace the existing six FFG-7 type escorts of the *Santa Maria* class. In October 2015, the Spanish Council of Ministers authorised the project definition stage for the class. Contracts worth c. €165m (US$180m) have subsequently been awarded for the design of the ships, their combat systems and key sensors.[8] The new ships are expected to be a little smaller than the existing F-100 class *Álvaro de Bazán* class air-defence frigates and be equipped with multi-mission capabilities. Key design features include an inte-

Table 2.4.4: SPANISH NAVY: PRINCIPAL UNITS AS AT MID 2016

TYPE	CLASS	NUMBER	TONNAGE	DIMENSIONS	PROPULSION	CREW	DATE
Principal Surface Escorts							
Frigate – FFG	**ÁLVARO DE BAZÁN** (F-100)	5	6,300 tons	147m x 19m x 5m	CODOG, 28 knots	200	2002
Frigate – FFG	**SANTA MARIA** (FFG-7)	6	4,100 tons	138m x 14m x 5m	COGAG, 30 knots	225	1986
Submarines							
Submarine – SSK	**GALERNA** (S-70/AGOSTA)	3	1,800 tons	68m x 7m x 6m	Diesel-electric, 21 knots	60	1983
Major Amphibious Units							
Amph Assault Ship – LHD	**JUAN CARLOS I**	1	27,100 tons	231m x 32m x 7m	IEP, 21 knots	245	2010
Landing Platform Dock – LPD	**GALICIA**	2	13,000 tons	160m x 25m x 6m	Diesel, 20 knots	185	1998

grated mast, a combined diesel electric and gas (CODLAG) propulsion system and a flexible mission bay. Meanwhile, continuity of production at domestic shipbuilder Navantia is being maintained by the orders for two additional *Meteoro* class vessels ordered in December 2014. The keels of both new vessels were laid at simultaneous ceremonies at Ferrol in Galicia and San Fernando, Cadiz on 29 April 2016. Launch will take place in the second half of 2017. One ship – to be named *Audaz* – will be delivered by June 2018 and the other – *Furor* – will follow by year-end. The navy still hopes to extend the class to a total of ten ships by an order for a third batch of four units, including two specialised hydrographic/submarine rescue variants.

Meanwhile work is slowly progressing on the badly-delayed S-80 submarine programme following the emergence of design problems relating both to their buoyancy and the poor performance of their new air-independent propulsion (AIP) system. The renamed S-81 Plus project is focused on completing the second boat in the series, *Narcisco Monturiol*. She has been lengthened following completion of a c. 10m hull plug at the end of 2015. A critical design review due for conclusion during the second half of 2016 is set to finalise the pathway through to her delivery, which is now scheduled for the end of 2021. Work on the other three submarines in the class has been suspended until the revised design has been fully certified. The navy is planning to extend the lives of its three remaining *Galerna* ('Agosta') class submarines to compensate for the c. nine-year delay on the S-80/S-81 Plus project through completion of a fifth cycle of major overhauls. This may not be an easy undertaking given the class first entered service as long ago as 1983.

A looming problem for the navy is how to replace its existing small force of EAV-8B Harrier Plus jets, which will approach life-expiry a decade or so from now. The F-35B vertical/short take-off and landing (V/STOL) variant of the Joint Strike Fighter is seen to be the only possible replacement but funding is a problem. The navy hopes to collaborate with the air force on a joint F-35 purchase but their cooperation is not ensured.

There have been no significant changes year-on-year to the force structure set out in Table 2.4.4.

Spanish shipbuilder Navantia is beginning to receive new orders after a difficult period during the Eurozone crisis. Two replenishment vessels similar to *Cantabria* – seen in the centre of this photograph – have been ordered by the Royal Australian Navy. In addition, design work is underway on the new F-110 class frigates that will replace the Spanish Navy's existing FFG-7 type, including *Santa Maria* pictured right. *(Navantia)*

MAJOR REGIONAL POWERS – UNITED KINGDOM

The mid-term direction of the British Royal Navy has been determined by the 2015 Strategic Security and Defence Review (SDSR), which was published on 23 November 2015.[9] Taking a five- to ten-year planning horizon, the review aimed to prioritise the allocation of resources across the defence, security and diplomatic services to counter the greatest risks posed to national security. In marked contrast to

Table 2.4.5: BRITISH ROYAL NAVY: PRINCIPAL UNITS AS AT MID 2016

TYPE	CLASS	NUMBER	TONNAGE	DIMENSIONS	PROPULSION	CREW	DATE
Principal Surface Escorts							
Destroyer – DDG	**DARING** (Type 45)	6	7,500 tons	152m x 21m x 5m	IEP, 30 knots	190	2008
Frigate – FFG	**NORFOLK** (Type 23)	13	4,900 tons	133m x 16m x 5m	CODLAG, 30 knots	185	1990
Submarines							
Submarine – SSBN	**VANGUARD**	4	16,000 tons	150m x 13m x 12m	Nuclear, 25+ knots	135	1993
Submarine – SSN	**ASTUTE**	3	7,800 tons	93m x 11m x 10m	Nuclear, 30+ knots	100	2010
Submarine – SSN	**TRAFALGAR**	4	5,200 tons	85m x 10m x 10m	Nuclear, 30+ knots	130	1983
Major Amphibious Units							
Helicopter Carrier – LPH	**OCEAN**	1	22,500 tons	203m x 35m x 7m	Diesel, 18 knots	490	1998
Landing Platform Dock – LPD	**ALBION**	2	18,500 tons	176m x 29m x 7m	IEP, 18 knots	325	2003
Landing Ship Dock – LSD (A)	**LARGS BAY**	3	16,200 tons	176m x 26m x 6m	Diesel-electric, 18 knots	60	2006

Notes:
1. Reports suggest that one Type 45 destroyer and one Type 23 frigate are effectively in extended readiness.

most other British defence reviews published since the end of the Cold War, SDSR 2015 responded to the changed international environment by providing for a modest uplift in defence funding and force numbers. This was reflected in an enlarged 'Joint Force 2025' vision that envisages up to 50,000 personnel – including a maritime task group centred on a *Queen Elizabeth* class aircraft carrier – being employed on an expeditionary operation.

From a naval perspective, SDSR 2015 was broadly positive in spite of a number of disappointments. The British Admiralty had prioritised three core capabilities of (i) continuous at-sea deterrence, (ii) carrier strike and (iii) amphibious readiness in advance of the review. The means to ensure these were largely delivered.[10] Whilst headline announcements tended to focus on bolstering air power, many of these decisions – notably the acquisition of Boeing P-8 maritime patrol aircraft and an acceleration of orders for F-35B Joint Strike Fighters – were also key 'enablers' of the Admiralty's own objectives. The review was also significant in approving a modest increase in trained naval manpower to c. 30,500 by 2020. This is around 1,500 more than the level planned in the previous SDSR 2010 and should alleviate recent manning issues. Major shipbuilding programmes were largely protected and work on a new National Shipbuilding Strategy should ensure better continuity of protection in the longer term. The current position with respect to key warship types is summarised in greater detail below, whilst Table 2.4.5 shows the fleet's main current constituents.

Aircraft Carriers and Amphibious Ships: SDSR 2015 affirmed previous decisions that both *Queen Elizabeth* class carriers would be brought into service. Equally significantly, both will be manned so as to ensure that one is continuously available. Initial sea trials and delivery of the lead ship are now expected to take place early in 2017 prior to an extended period of work-up. This will culminate in an operational carrier strike capability being declared by the end of 2020. The concept of facilitating the use of the ships in a number of scenarios through the deployment of tailored air groups remains a high priority and at least one of the class – initially *Prince of Wales* – will be modified to increase amphibious warfare capabilities. *Prince of Wales* is now structurally complete and her float-out will take place in 2017 before delivery during 2019.

In spite of attempts to gain a stay of execution, the helicopter carrier *Ocean* will retire, as planned, during 2018. The two amphibious transport docks will remain in the fleet, with – as currently – one being held at extended readiness. *Albion* is currently being refitted and modernised at Devonport to assume the role of operational unit. She will commence sea trials in 2017 prior to replacing *Bulwark* in the active fleet. SDSR 2015 also confirmed a longer term requirement as part of Future Force 2025 for the three remaining auxiliary dock landing ships of the 'Bay' class.

Surface Combatants: SDSR 2015 retained the fleet of manned surface combatants at a minimum of nineteen vessels but changed its planned composition. It had become increasingly apparent during the course of 2015 that the targeted fleet of thirteen Type 26 frigates had become increasingly unaffordable. Some media reports suggested that planned costs were approaching around £1bn (US$1.4bn) per unit or more than double that initially envisaged.[11] It was therefore determined to reduce the planned Type 26 fleet to just eight anti-submarine focused vessels and acquire a lighter, cheaper frigate for general-purpose tasking. The new ship will be known as the Type 31 and – alongside the Type 26 – will form a key component of the National Shipbuilding Strategy set to be released at the end of 2016. At least five, possibly more, will be built.

At present, there is considerable uncertainty as to what form the Type 31 design will take, although variants of BAE System's *Al Shamikh* class (built for the Royal Navy of Oman) and BMT Defence Services' 'Venator 110' concept appear to be under consideration. It seems that that a degree of dissatisfaction with BAE Systems' management of previous shipbuilding projects and lack of export successes could result in a greater role in the Type 31 programme being allocated to other British maritime companies. This may include at least some fabrication taking place away from the Clyde. Another area of uncertainty relates to the scheduling of construction of both types. The start of Type 26 fabrication has been continually postponed. Whilst this has been rather unconvincingly explained away on the basis of ensuring greater design maturity, money appears to be the real issue. Given the pressing need to start replacing the existing Type 23 frigates early in the

A computer-generated image of the latest iteration of the new Type 26 frigate design. Its cost is significantly higher than originally estimated and the programme was cut from thirteen to eight by SDSR 2015. Instead, it is planned to supplement the class with orders for a lighter and cheaper frigate design, which will be known as the Type 31. *(BAE Systems)*

BMT Defence Service's 'Venator 110' concept design is one contender for the new Type 31 light frigate, also increasingly referred to as the General Purpose Frigate. *(BMT Defence Services)*

Pending the start of work on the new 'Successor' class strategic submarine, British submarine construction is focused on the ongoing *Astute* class programme. The third boat in the class, *Artful*, commenced sea trials in August 2015 and was commissioned on 18 March 2016. *(BAE Systems)*

2020s, there are even some suggestions that the initial Type 31 frigates may be completed before Type 26 production gets fully underway.

One reason for the strained relationship between BAE Systems and its major customer has been the emergence of significant problems with the innovative integrated electronic propulsion system used in the Type 45 *Daring* class destroyers. The reasons for the problems are complex but appear to relate largely to greater-than-expected unreliability with the inter-cooled recuperated (ICR) technology used with the ship's WR-21 gas turbines coupled with a mismatch in power output between these turbines and the much smaller diesel generators that supply the ships' electrical grid. This can lead to a complete loss of propulsion if one of the turbines fails. The problem can be exacerbated in extremely hot conditions such as the Persian Gulf, where the class has been frequently deployed.[12] Significant enhancements to system reliability have already been achieved under the so-called Project Napier and SDSR 2015 has allocated funding to ensure greater resilience by bolstering the installed output provided by diesel generation. However, this will take time to implement. In the meantime, the problems have generated considerable negative publicity for what has otherwise proved to be a world-beating design.

Minor Warships and Auxiliaries: The delays to the Type 26 programme have already resulted in three Batch 2 'River' class offshore patrol vessels being ordered to ensure continuity of production at BAE Systems' yards on the Clyde. SDSR 2015 confirmed plans to order two additional vessels. They are likely to be used to cover additional low-intensity taskings traditionally undertaken by major surface combatants. Work on the class is progressing slowly, with the first ship – *Forth* – expected to be launched in the summer of 2016. The new ships will replace the existing, smaller original 'River' class vessels of the *Tyne* class but the future of the Falkland Islands patrol vessel *Clyde* has yet to be determined.

The numbers of mine countermeasures vessels in service are expected to fall to twelve units by 2025 as older *Sandown* class vessels are retired. Preference is being given to retaining the older 'Hunt' class in service as their larger size better supports new unmanned capabilities now being developed, most notably under the Maritime Mine Counter Measures (MMCM) programme being pursued jointly with France.

A major – if little reported – project funded in SDSR 2015 was that for three new solid support ships. These will provide vital replenishment and logistic capabilities for the new carriers and therefore represent an important element of the Royal Navy's future vision. It seems likely – subject to the outcome of the National Shipbuilding Strategy – that the new ships will be built abroad in the same fashion as the current 'Tide' class. However, it is interesting to note that deliveries of these vessels from Korea's DSME are running significantly behind schedule, demonstrating that it is not only the British shipbuilding sector that can suffer from production issues.

Submarines: Current submarine production is dominated by ongoing construction of the *Astute* class. The third boat, *Artful*, was commissioned on 18 March 2016 and a formal main build order for fifth of class, *Anson*, placed the previous November. Work has been underway on her from as long ago as 2010 and fabrication of the remaining two boats is also well advanced. Unsurprisingly, SDSR 2015 left the scope of the follow-on project for 'Successor' strategic submarines largely unchanged. However, projected cost was revised upwards to £31bn (US$43bn) plus a £10bn (US$14bn) contingency over the life of a twenty-year programme. It also seems that the first submarine will not be delivered until the early 2030s. This is later than previously envisaged and will put a strain on the existing, ageing boats. Investment in design work and long lead items for the class already amounts to c. £5bn (US$7bn) but the formal construction phase has yet to be announced. Its timing may be impacted by the political turmoil caused by the United Kingdom's 23 June 2016 decision to leave the European Union.

Operationally, the strain on the Royal Navy's depleted front-line forces has been demonstrated by the 'gapping' of certain standing commitments to allow warships to be released for increasingly important NATO taskings. Offshore patrol vessels and auxiliaries have also been allocated to duties previously assigned to frigates and destroyers. Achieving full manning levels also seems to remain problematic. Two surface combatants – the Type 45 destroyer *Dauntless* and the Type 23 frigate *Lancaster* – are reportedly being held at reduced readiness in spite of the real pressure on the surface fleet.[13] The Middle East, however, remains a focal point for Royal Navy operations and resources, as evidenced by work to establish the new permanent base at Mina Salman in Bahrain.

MID-SIZED REGIONAL FLEETS

Germany: German defence spending is on a markedly rising trend as the government scrambles to adjust to the potential Russian threat to the east. Proposed defence spending is set to rise from €34.3bn (c. US$38bn) in 2016 to €39.2bn (US$43bn) in 2020 as part of moves to fund a €130bn (US$145bn), fifteen-year procurement programme and increase overall numbers of personnel. Whilst the future direction of defence strategy will be clarified by a new defence white paper due to be released before the end of 2016, it is already apparent that the army will be the main beneficiary of the planned largesse.

Indeed, naval force levels are still shrinking as previously-planned reductions feed through to operational units. The last year has seen the withdrawal of three further minehunters and two of the depleted fleet of fast attack craft. The remaining four Type 143A *Gepard* class vessels will be withdrawn before the end of 2016. Conversely, no new major warships have joined the fleet over the past twelve months. The final Type 212A submarine *U36* had yet to be delivered as of June 2016 and the lead F-125 stabilisation frigate, *Baden-Württemberg*, is in the early stages of sea trials.

Baden-Württemberg's maiden voyage on 7 April 2016 was a welcome development for a programme that has been marked by considerable delays and cost overruns. It is now hoped to deliver the new ship during 2017 and that all four ships in the class will be operational before 2020. The third unit, *Sachsen-Anhalt*, was christened in March 2016 and the fourth should follow over the next year. Production will than transition to the new MKS-180 surface combatant, which is steadily increasing in size and cost. An order for four ships – plus two options – is expected during 2017. The new ships will have much greater warfighting potential than the stabilisation-orientated F-125 design. However, a likely displacement of around 8,000 tons still leaves them optimised for longer-range deployment at a time when attention is returning to the Baltic.

Greece: The Hellenic Navy continues to make headway with completing programmes effectively suspended during its economic crisis. This has been

Baden-Württemberg, the first of the new German F-125 class of stabilisation frigates, commenced trials on 7 April 2016, somewhat later than initially envisaged. This picture shows her departing Hamburg on her maiden voyage. *(ThyssenKrupp Marine Systems)*

reflected in the long-delayed delivery of the fifth *Roussen* class fast attack craft, *Ritsos*, in a ceremony on 19 October 2015 and the effective completion of the Type 214 submarine project. A commissioning ceremony for the final two submarines – *Matrozos* and *Katsonis* – in a four-boat programme was held on 23 June 2016. Previous plans called for completion of a further two submarines of the class but this seems unlikely given Greece's ongoing economic weakness. Indeed, delivery of the new boats should allow withdrawals of older Type 209 submarines, all of which date from the 1970s.

Meanwhile, 2015 saw the Aegean become a new fulcrum for the European migration crisis, overtaking the previous sea route between Libya and North Africa as the key entry-point into the European Union (EU). According to some estimates, around eighty percent of more than one million migrants entering the EU in 2015 arrived in Greece by sea, largely making the short but dangerous journey from Turkey to the offshore Greek islands.[14] The crisis saw NATO deploy maritime forces to help the EU's border management agency Frontex cope with a humanitarian catastrophe which has cost hundreds of lives. However, it is a deal with Turkey to take back asylum seekers using illegal routes to the EU that is reportedly having most effect, dramatically reducing the number of migrants attempting the crossing. Unfortunately, this may well result in migrant attention returning to the central Mediterranean, where even more have died.

The Netherlands: A full update on the major influences impacting the Royal Netherlands Navy's current objectives and structure is provided in Chapter 2.4B. Major recent developments include completion of the mid-life refit of the *Walrus* class submarine *Zeeleeuw* towards the end of 2015. She is the first of the four boats in the class to be modernised under a programme that will extend service life by around ten years. More negatively, the new joint support ship *Karel Doorman* suffered a major failure of one of her electrical propulsion motors in March 2016. She is likely to be out of operation for around eight months until a replacement is installed.

Turkey: An indication of the extent of Turkey's expanding naval ambitions was provided on 30 April 2016 at a ceremony to mark the start of construction of the new amphibious assault ship *Anadolu* at the Sedef Shipyard, Istanbul. During a speech at the ceremony, Turkey's President Recep Tayyip Erdoğan confirmed that STOVL F-35B Joint Strike Fighters would be acquired to operate from the ship. He also hinted that fully-fledged aircraft carriers may be constructed in due course. The new ship is scheduled to be delivered by 2021 but President Erdoğan

The Turkish Navy's indigenous construction programmes are moving slowly forwards. This view shows the third 'Milgem' or 'Ada' class corvette, *Burgazada*, shortly before she was launched in May 2016. *(Devrim Yaylali)*

hopes that this can be shortened. *Anadolu* is a half-sister of the Spanish *Juan Carlos I* design, having similar overall dimensions but incorporating an all-diesel power generation plant and Turkish manufactured combat systems and sensors.[15] She is at the heart of a renewal of the navy's amphibious forces that has already seen the delivery of eight small tank landing craft and which also encompasses two larger tank landing ships. One of these, *Bayraktar*, was launched in September 2015. This cleared the slipway for work to commence on the latter ship, which will be named *Sancaktar*.

Unfortunately, other Turkish Navy programmes are not moving ahead quite so quickly due to a mix of contractual, political and technical complications. June 2016 saw the launch of *Burgazada*, the third 'Milgem' or 'Ada' type corvette of the *Heybeliada* class. The fourth ship, *Kinaliada*, was laid down on the same day. This represents a gap of nearly five years since the launch of the second ship, *Büyükada*. Moreover, work has yet to commence on a second batch of four 'I' class variants, which are due to be equipped with the Mk.41 vertical launch system. The start of construction on the first of these, to be named *Istanbul*, is targeted for January 2017. The delays are having a knock-on effect on other surface programmes, most notably the TF-2000 air-defence frigates that are needed to modernise fleet air-defence capabilities.

There have also been delays with the local build of six AIP-equipped Type 214 submarines. A contract for their acquisition was signed with ThyssenKrupp Marine Systems as long ago as July 2009 but it was only in October 2015 that work on the first boat, *Piri Reis*, finally commenced. Devising a suitable method of integrating locally-manufactured sub-systems with the core German-built combat system might be part of the reason for this delay. It is now hoped to deliver *Piri Reis* during 2021, some six years later than initially planned.

OTHER REGIONAL FLEETS

Black Sea and Mediterranean: The Black Sea is currently one of the front-lines in the new Cold War between NATO and Russia. However, local procurement activity remains muted. **Ukraine** has big plans to rebuild a fleet depleted by the loss of ships and bases due to Russia's annexation of the Crimea but it is difficult to imagine how these will be funded. The most tangible recent announcement has therefore come from **Bulgaria**. It aims to buy two new

The Romanian Navy is hoping to commence work on the second phase of a modernisation programme planned for its two former British Royal Navy Type 22 frigates. This view of *Regina Maria* (formerly *London*) off Constanta in August 2014 suggests she is currently relatively lightly armed compared with her Royal Navy days. *(US Navy)*

The Irish Naval Service is steadily taking delivery of new PV90 type offshore patrol vessels of the *Samuel Beckett* class. This image shows the second member of the class, *James Joyce*, which was delivered in July 2015 and commissioned on 1 September of that year. A third ship, *William Butler Yeats*, is about to start sea trials and a fourth has been ordered. *(Irish Defence Forces)*

offshore patrol ships at a cost of €420m (US$460m) as part of a larger €1.25bn package aimed at bolstering the navy and air force. This figure suggests the purchase of sophisticated ships, perhaps a variant of DCNS' 'Gowind' series that came close to being acquired previously before the global recession hit. Plans to modernise some of the existing *Wielingen* class frigates bought second-hand from Belgium will also go ahead. Meanwhile, **Romania** hopes to restart its delayed programme to undertake a second phase upgrade of its two former British Type 22 frigates that were originally purchased in 2003.

Elsewhere in the region, **Croatia** and **Montenegro** have both been undertaking limited modernisation programmes aimed more at coast guard than war-fighting functions. The impact of the migration crisis on **Malta** is also leading to patrol forces being bolstered. In June 2016, the European Union allocated funds to support the acquisition of a new €34m (US$37.5m) patrol vessel. The existing *Diciotti* type patrol vessel *P61* is also being upgraded to improve her capability.

Atlantic and North Sea: One potential bidder for the new Maltese patrol ship is Babcock Marine's shipyard in Appledore, North Devon. It has been steadily building a series of offshore patrol vessels for **Ireland**, delivering a second *Samuel Beckett* class ship, *James Joyce*, in July 2015. A third member of the class, *William Butler Yeats*, is due to start sea trials in July 2016, replacing the recently-decommissioned *Aisling*. In addition, an unexpected order for a fourth vessel was placed the previous month at a cost of some €54m (US$60m). A new Irish defence white paper published in August 2015 committed – as a minimum – to maintaining the existing Irish Naval Service fleet of eight vessels but it seems that an aspiration of a larger force may now be met.

Portugal's naval structure continues to maintain a balance between a small force of front-line frigates and submarines supplemented by larger numbers of constabulary assets of various sizes. Most of the latter are elderly and long overdue for replacement. The confirmation of an order for a second pair of *Viana do Castelo* class offshore patrol ships in July 2015 was therefore a positive development. Further good news was provided by the commissioning of *Tejo*, the first of four smaller Stanflex 300-type vessels acquired from Denmark, on 5 May 2016.

The security agenda in **Belgium** has been dominated by terrorist actions in Brussels. However, an event of longer-term strategic importance was the agreement of a new strategic plan for 2030, adopted in December 2015. This essentially entails a reduction in overall manpower in return for a major boost to procurement. For the Belgian Naval Component, a commitment to maintain a force of two frigates and six mine countermeasures vessels retains the current fleet structure basically intact.[16] It would seem both future requirements will be pursued in conjunction with the Netherlands, which will need to replace its own largely identical ships in a similar timescale.

Scandinavia and the Baltic: The Baltic has also become a focal point of the increased tension between Russia and the West. This is leading to a significant rebalancing of naval forces back towards territorial defence, leading to a renewed emphasis on assets such as submarines and war-fighting vessels. Leading the way is **Sweden**, which concluded a contract for the construction of two new A26 class submarines on 30 June 2015. The deal, which also encompasses the upgrade of two of the existing *Gotland* class boats and an option on the third, brings expenditure on modernisation of underwater forces to nearly SEK10bn (US$1.2bn). It is part of a series of measures across the armed forces that includes stronger air-defence, anti-submarine and mine-warfare capabilities, as well as re-establishment of a permanent military presence on the island of Gotland.[17] Fabrication of the first new submarine commenced soon after the contract was awarded, with the first steel cut on 4 September 2015.

Also placing an emphasis on submarine procurement are neighbouring **Norway** and **Poland**. Much to Sweden's disappointment, the former has short-listed German and French designs for its new class of submarines, which will replace the six-strong *Ula* class. A firm decision to proceed with the programme has yet to be taken but this seems all-but inevitable given the current geopolitical backdrop. Meanwhile Poland, which plans to buy three submarines as part of a PLN13bn (US$3bn) fleet

An order for two new A26 class submarines for the Royal Swedish Navy was signed on 30 June 2015. An interesting feature of the design is a multi-mission portal, similar in concept to the flexible mission decks increasingly found on surface warships, located between the four torpedo tubes. This can be used to launch and retrieve a diverse range of payloads, such as unmanned vehicles or Special Forces. *(Copyright Saab AB)*

The Norwegian *Ula* class submarine *Utsira* pictured on exercises in the Norwegian fjords in 2015. Norway is one of a number of Scandinavian and Baltic navies pondering how to renew their underwater forces. *(Petter Brenni Gulbrandsen/Norwegian Armed Forces)*

modernisation plan through to 2030, has yet to select potential suppliers.

Finland's main naval procurement effort remains its 'Squadron 2020' programme for new surface combatants to replace its four *Rauma* class fast attack craft and remaining minelayers. At least four missile-armed units with an ability to lay mines are likely to be acquired. The planning phase of the €1.2bn (US$1.3bn) project will last until 2018, being followed by the start of construction the following year to ensure the start of deliveries during the early 2020s. There were reports in the Finnish press during 2015 that fatigue damage had been found in the *Rauma* class after their recent mid-life upgrades, suggesting replacement is increasingly pressing.

RUSSIA

The dramatic increase in Russian defence spending in recent years has spawned a huge rise in naval construction that a weakened shipbuilding and armaments base has struggled to cope with. Multiple problems have included reliance on component suppliers that – in a legacy of former Soviet policy – are dispersed across the constituents of the former Soviet Union, difficulties in integrating a raft of new equipment and a simple lack of construction capacity. The interruption of deliveries of key equipment – notably Ukrainian gas turbines and German diesel engines as a result of the Crimean intervention – has exacerbated these problems. A particular feature of recent construction has been the authorisation of programmes for an almost bewildering range of warships, particularly in terms of smaller surface combatants.[18] This practice may also be a negative legacy of the former Soviet era, when numerous state design bureaux effectively competed for work. In any event, the result would appear to be the procurement of uneconomic numbers of different classes, often incorporating different combat systems and other equipment. The problem of sustaining these multiple types is likely to become all the more problematic now that Russia's declining economy is putting defence spending under renewed pressure.

Table 2.4.6 summaries current major warship classes, whilst the following summary attempts to untangle the web of current procurement.

Submarines: Submarine construction remains the Russian Navy's top strategic priority. Possibly because of its complexity, it is also the area where there is greater standardisation. Strategic submarine assembly if firmly focused on the Project 955A variant of the 'Borey' type at the Sevmash plant in Severodvinsk. Four of these boats are now under construction following a keel-laying ceremony for *Imperator Aleksandr III* in December 2015. Work on a fifth is also expected to start in the second half of 2016. Together with the three earlier Project 955 variants, this will allow a long-stated objective of having eight members of the class in operational service by the early 2020s to be achieved. The second Project 955 boat, *Aleksandr Nevsky*, has already deployed to the Far East as part of plans to renew the Pacific Fleet's strategic submarine flotilla. She will be joined by *Vladimir Monomakh* before the end of 2016. However, there are conflicting reports on the success or otherwise of the latter's latest test firing of two of her RSM-56 Bulava strategic missiles in November 2015. It appears a further trial is planned before she departs to her new base.

TABLE 2.4.6: RUSSIAN NAVY: SELECTED PRINCIPAL UNITS AS AT MID 2016

TYPE	CLASS	NUMBER[1]	TONNAGE	DIMENSIONS	PROPULSION	CREW	DATE
Aircraft carriers							
Aircraft Carrier – CV	Project 1143.5 **KUZNETSOV**	1	60,000 tons	306m x 35/73m x 10m	Steam, 32 knots	2,600	1991
Principal Surface Escorts							
Battlecruiser – BCGN	Project 1144.2 **KIROV**	1 (1)	25,000 tons	252m x 29m x 9m	CONAS, 32 knots	740	1980
Cruiser – CG	Project 1164 **MOSKVA** ('Slava')	3	12,500 tons	186m x 21m x 8m	COGAG, 32 knots	530	1982
Destroyer – DDG	Project 956/956A **SOVREMENNY**	c.5	8,000 tons	156m x 17m x 6m	Steam, 32 knots	300	1980
Destroyer – DDG	Project 1155.1 **CHABANENKO** ('Udaloy II')	1	9,000 tons	163m x 19m x 6m	COGAG, 29 knots	250	1999
Destroyer – DDG	Project 1155 **UDALOY**	c.8	8.500 tons	163m x 19m x 6m	COGAG, 30 knots	300	1980
Frigate – FFG	Project 1136.6M **GRIGOROVICH**	2	4,000 tons	125m x 15m x 4m	COGAG, 30 knots	200	2016
Frigate – FFG	Project 1154 **NEUSTRASHIMY**	2	4,400 tons	139m x 16m x 6m	COGAG, 30 knots	210	1993
Frigate – FFG	Project 1135 **BDITELNNY** ('Krivak I/II')	c.2	3,700 tons	123m x 14m x 5m	COGAG, 32 knots	180	1970
Frigate – FFG	Project 2038.0 **STEREGUSHCHY**	4	2,200 tons	105m x 11m x 4m	CODAD, 27 knots[2]	100	2008
Frigate – FFG	Project 1161.1 **TATARSTAN** ('Gepard')	2	2,000 tons	102m x 13m x 4m	CODOG, 27 knots	100	2002
Submarines							
Submarine – SSBN	Project 955 **YURY DOLGORUKY** ('Borey')	3	17,000+ tons	170m x 13m x 10m	Nuclear, 25+ knots	110	2010
Submarine – SSBN	Project 941 **DONSKOY** ('Typhoon')	1	33,000 tons	173m x 23m x 12m	Nuclear, 26 knots	150	1981
Submarine – SSBN	Project 677BDRM **VERKHOTURYE** ('Delta IV')	6	18,000 tons	167m x 12m x 9m	Nuclear, 24 knots	130	1985
Submarine – SSBN	Project 677BDR **ZVEZDA** ('Delta III')	3	12,000 tons	160m x 12m x 9m	Nuclear, 24 knots	130	1976
Submarine – SSGN	Project 855 **SEVERODVINSK** ('Yasen')	1	13,500 tons+	120m x 14m x 9m	Nuclear, 30+ knots	90	2013
Submarine – SSGN	Project 949A ('Oscar II')	c.5	17,500 tons	154m x 8m x 9m	Nuclear, 30+ knots	100	1986
Submarine – SSN	Project 971 ('Akula I/II')	c.10	9,500 tons	110m x 14m x 10m	Nuclear, 30+ knots	60	1986
Submarine – SSK	Project 677 **ST PETERSBURG** ('Lada')	1	2,700 tons	72m x 7m x 7m	Diesel-electric, 21 knots	40	2010
Submarine – SSK	Project 877/636 ('Kilo')	c.20	3,000 tons	73m x 10m x 7m	Diesel-electric, 20 knots	55	1981

Notes:
1 Table only includes main types and focuses on operational units: bracketed figures are ships being refurbished or in maintained reserve.
2 Some sources state CODOG propulsion.

Ivan Gren, the first of a new class of large landing ships being built for the Russian Navy pictured departing the Yantar shipyard in Kaliningrad for sea trials. She is scheduled for delivery before the end of 2016. *(Yantar)*

The Project 1135.6 *Admiral Grigorovich* class frigates are sisters to the *Talwar* class already in service with the Indian Navy. They were ordered to bolster fleet numbers due to delays in the construction of more modern designs. Six have been ordered from the Yantar yard but work on the last three has been suspended because their Ukrainian-built gas turbines have not been delivered. This picture shows second of class, *Admiral Essen*, in April 2016, shortly before her delivery to the Russian Navy. *(Yantar)*

Sevmash is also responsible for the sole current nuclear-powered attack submarine programme, which involves the construction of the Project 885M variant of the 'Yasen' class. Four production variants are now in the course of assembly following delivery of the prototype Project 885 boat *Severodvinsk* in 2014 and a fifth will be laid down shortly. Total Project 855/855M production could amount to seven submarines or more.

The third strand of Russian submarine production was intended to focus on construction of Project 677 diesel-electric 'Lada' class boats at the Admiralty Shipyard in St. Petersburg. Only the lead submarine, *St. Peterburg*, has been commissioned to date and her performance has been disappointing. Completion of two other members of the class has been continuously pushed back and plans to equip them with AIP seem to have been abandoned.[19] The problems have resulted in the construction of an upgraded Project 636.3 variant of the Cold War 'Kilo' class type at the Admiralty yard. Six of these were ordered for the Black Sea fleet, of which four had been commissioned by June 2016. At a ceremony to mark the roll-out of the final boat, *Kolpino*, on 31 May 2016, it was confirmed that another batch of the submarines would be ordered to reinforce Russia's Pacific Fleet.

Aircraft Carriers and Amphibious Ships: With hopes for a new generation of aircraft carriers still some years from fruition, long-mooted plans to undertake a major modernisation of the sole existing ship, the Project 1143.5 *Admiral Kuznetsov*, appear to have been revived. Work is likely to commence after a planned Mediterranean deployment in early 2017 and could take up to three years to complete. As for France, the lack of an available aircraft carrier during her absence highlights the risks inherent in a one carrier-strategy.

Turning to amphibious ships, sea trials of the prototype Project 1171.1 landing ship *Ivan Gren* have now commenced. If these are successful, she will be transferred to the fleet during the autumn of 2016. A second ship, *Petr Morgunov*, is under construction for delivery in 2018.

Surface Vessels: It is in the area of surface warship construction that Russian shipbuilding efforts appear to suffer from most duplication. There are at least seven ongoing programmes, each of which is summarised below:

Project 2235.0 – *Admiral Gorshkov* Class: These frigates are intended to be the Russian Navy's main class of blue water surface combatants in the medium term. They displace c. 4,500 tons in full load condition and incorporate the new Poliment 5P-20K active phased array and a balanced outfit of weaponry. The lead ship was laid down in 2006 and is currently completing an extended series of sea trials. It should be delivered soon. Three more are under construction but two will be delayed until an indigenous replacement for their Ukrainian-manufactured turbines can be supplied. A total of up to twenty is planned. The ships are built by the Severnaya Verf shipyard in St. Petersburg to a Severnoye Design Bureau specification.

Project 1135.6 – *Admiral Grigorovich* Class: These frigates are an interim 4,000-ton design ordered due to delays with other programmes. They are close sisters to India's *Talwar* class, themselves a significantly updated version of the Cold War 'Krivak' series of anti-submarine escorts. The first two – *Admiral Grigorovich* and *Admiral Essen* – were commissioned in the first half of 2016. The third, *Admiral Markarov*, will start sea trials soon. Another three have suffered similar problems from the unavailability of Ukrainian turbines as the later members of the Project 2235.0 type. The class is also built to a Severnoye Design Bureau specification but is being constructed at the Yantar shipyard in Kaliningrad.

Project 2038.0/2038.5 – *Steregushchy/Gremyashchy* Classes: The lead Project 2038.0 class ship, first of a series of 2,200-ton littoral warfare frigates/corvettes, was laid down in 2001 and commissioned towards the end of 2007. Three further ships have been completed and six more are being built by two separate production lines at Severnaya Verf and at the Far Eastern Komsomolsk Shipyard in Komsomolsk-on-Amur. The upgraded and larger Project 2038.5 variant incorporates a different VLS arrangement and was to be equipped with German MTU diesels. These have now been replaced by a Russian engine but it appears construction beyond the first two vessels has now been suspended in favour of the earlier type. Both variants were designed by the Almaz Central Design Bureau.

The Russian Navy has been constructing quite large numbers of small missile-armed corvettes which are restricted to relatively sheltered waters but pack a powerful punch. Amongst these are the Project 2163.0/2163.1'Buyan/Buyan M' series of artillery and missile ships – *Serpukhov* of the Project 2163.1 variant is shown here whilst transiting the Bosphorus in March 2016. Three of the class were used to conduct Kalibr cruise-missile strikes on Syrian rebels during Russia's intervention in the civil war there. *(Devrim Yaylali)*

Project 2163.0/Project 2163.1 – 'Buyan'/'Buyan M': Officially designated as small artillery and small missile ships respectively, these are compact corvette-like ships that displace less than 1,000 tons. They were originally designed by the Zelenodolsk Design Bureau for service in the Caspian Sea. The three original Project 2163.0 variants were built at the Almaz shipyard at St. Petersburg, commissioning between 2006 and 2012. Production of the improved Project 2163.1 class was then transferred to the Zelenodolsk yard in Tatarstan, which has delivered five and is building four more. This latter variant is fitted with a vertical launch system for Kalibr surface-to-surface missiles, which three of the class used to attack targets in Syria in October 2015.

Project 22800 – 'Karakurt': This is a new class of small, 800-ton corvettes that appears to be very similar to the 'Buyan' class in overall concept. A creation of the Almaz Central Design Bureau, the first pair were laid down at the Pella shipyard in St. Petersburg on 24 December 2015. A further unit was laid down at the Morye Shipyard in the Crimea in May 2016 and reports suggest a total class of at least eighteen is planned. Like the Project 2163.1 series, the class is armed with Kalibr surface-to-surface missiles. It also incorporates a phased-array radar.

Project 22160: Nominally a class of lightly-armed offshore patrol vessels, these c. 2,000-ton ships feature a modular arrangement that allows a wide range of weapons systems and sensors to be accommodated. Six have been ordered from the Zelenodolsk yard to a Severnoye Design Bureau concept and four of these are currently under construction.

Project 23550: Known as Arctic corvettes or patrol ships, two of these hybrid vessels – combining the qualities of a tug, an ice-breaker and a patrol vessel – were ordered from the Admiralty Shipyard in May 2016. They are the latest in a series of ice-capable ships that also include two Project 03182 Arctic support ships due for delivery from 2017 onwards. Other support vessels, as well as diesel-electric and nuclear-powered ice breakers, are in service with or on order for other Russian government agencies.

New construction continues to be accompanied by efforts to refurbish existing Soviet-era designs to squeeze a few more years of life out of the current fleet before new ships arrive in significant numbers. Although there is growing disquiet over the inefficiency of this process, work on the Project 1144 *Kirov* class battlecruiser *Admiral Nakhimov* is continuing as planned for completion by 2018. She will be followed into the Sevmash yard by *Pyotr Velikiy*, the only member of the class currently operational, which will undergo a similar modernisation lasting until 2021.

Notes

1. The new doctrine – termed the 'Russian Federation Marine Doctrine' – was reported on the President of Russia website on 26 July 2015. A copy of the report, which highlighted discussions relating to key aspects of the doctrine, can be found at: en.kremlin.ru/events/president/news/50060. A more detailed analysis of the doctrine – including its implications for force development – was provided in Nikolai Novichkov's 'Russian naval doctrine looks to the future', *Jane's Defence Weekly* – 19 August 2015 (Coulsdon: IHS Jane's, 2015), pp.24–5.

2. The quotation is from Russian Deputy President Dmitry Rogozin as reported in the 'Russian Federation Marine Doctrine' discussion of 26 July 2015.

3. There are contradictory views as to whether Aegis Ashore actually represents a threat to Russian strategic systems. The United States has consistently downplayed the system's potential. However, Russia is certainly anxious about the development of NATO ballistic missile defence capabilities. For example – as explained further in Chapter 2.4A – Russia's ambassador to Denmark was widely quoted as stating that Royal Danish Navy ships equipped with BMD tracking capabilities as part of a NATO 'missile shield' would be subject to targeting by Russian nuclear weapons.

4. Details of the basing plan for the B2M vessels were set out in 'Le troisième B2M sera finalement basé á La Réunion', *Mer et Marine* – 7 October 2015. The *Mer et Marine* website at www.meretmarine.com/fr remains an excellent source of news on French naval developments.

5. An English language translation of the *2015 White Paper for international security and defence* (Rome, Italian Ministry of Defence, 2015) can be found at: www.difesa.it/Primo_Piano/Documents/2015/07_Luglio/White%20book.pdf. A good summary of its conclusions is provided by Andrea Gilli, Alessandro R Ungaro & Alessandro Marrone in 'The Italian White Paper for International Security and Defence', *The RUSI Journal* – 160:6 (London: Royal Institute for Strategic Studies, 2015), pp.34–41.

6. A good example of criticism of the apparent changing scope of the vessels ordered under the Naval Law is provided by Enrico Piovesana's Italian language 'Marina militare, la "nave umanitaria" si transforma in portaerei,' (Italian Navy – the humanitarian vessel is transformed into an aircraft carrier) *Il Fatto Quotidiano* – 14 April 2016 (Rome: Il Fatto SpA, 2016). The article suggests that a second phase of the programme is envisaged to provide a further increase in Italian naval strength. However, the retirement of the influential Chief of Naval Staff and architect of the Naval Law, Admiral Giuseppe de Giorgi, in June 2016 may make further progress more difficult.

7. A full assessment of the current status of the PPA programme was provided by Tom Kington in 'Italy Reveals Innovations on New Naval Vessels', *Defense News* – 24 June 2016 (Springfield VA: Sightline Media Group, 2016).

8. For further information on the new frigate programme, see Nuria Fernández's 'Class F-110 New Generation Frigates for the Spanish Navy' in *European Security & Defence* – 3/2016 (Bonn: Mittler Report, 2016), pp.70–2. Ms Fernández is editor of the Spanish edition of the website infodefensa.com, itself a valuable source of information on Spanish Navy developments.

9. See *National Security Strategy and Strategic Security and Defence Review 2015: Cm 9161* (London: Her Majesty's Stationary Office, 2015).

10. The Royal Navy's objectives were set out by then First Sea Lord Sir George Zambellas in a speech at the Navy Club, London on 5 February 2016. A copy of his script can be found at: www.gov.uk/government/speeches/first-sea-lord-speech-on-sdsr-2015-and-the-royal-navy

11. Amongst a number of media reports on Type 26 programme costs, Andrew Chuter's 'Ballpark Cost Revealed For Royal Navy Frigates', *Defense News* – 25 September 2015 (Springfield VA: Sightline Media Group, 2015) is a particularly useful source of information. It quotes the Assistant Chief of Naval Staff (Ships) as attributing a £12bn cost to the then thirteen ship programme or c. £925m per vessel.

12. The problems with the Type 45 destroyer's propulsion system have been extensively covered but some of the published information is contradictory. Good sources of reliable information are the various records and documents arising from the House of Commons Defence Committee's Naval Procurement: Type 26 and Type 45 inquiry, which can currently be accessed at: www.parliament.uk/business/committees/committees-a-z/commons-select/defence-committee/inquiries/parliament-2015/royal-navy-type-26-and-45-16-17/

13. One of the many journalists to report current pressures on the British Royal Navy's surface fleet was Jamie Merrill in 'Falkland Islands left with no Royal Navy warship protection for the first time since 1982 conflict', *The Independent* – 1 April 2016 (London: Independent Digital News & Media, 2016).

14. Obtaining accurate figures on illegal migration to Europe is fraught with difficulty, partly because there is frequent double counting as migrants cross different borders. The figures quoted come from Tom Miles' 'EU gets one million migrants in 2015, smugglers seen making $1 billion', *Reuters* – 22 December 2015 (London: Reuters, 2015), an article itself largely based on International Organization for Migration (IOM) sources.

15. Devrim Yaylali's *Bosphorus Naval News* blog at turkishnavy.net remains an invaluable source of news on naval developments in Turkey and the surrounding area.

16. A good overview of Belgian defence plans was provided by Nathan Gain in 'Belgium adopts Strategic Defence Plan for 2030' on the *Forces Operations* blog forcesoperations.com/

17. The rebalancing of Scandinavian forces towards territorial defence was demonstrated in an interview by Kate Tringham of 'Peter Hultqvist, Minister of Defence, Sweden' in *Jane's Defence Weekly* – 25 November 2015 (Coulsdon: IHS Jane's, 2015), p.34 in which he was quoted as saying, 'National defence is number one for the Swedish defence forces right now, and we are planning for what we call national wartime scenarios'. Details of the Swedish orders were provided by Richard Scott's 'Sweden signs deal for A 26 submarine order, A 19 upgrade', *Jane's Defence Weekly* – 8 July 2015 (Coulsdon: IHS Jane's, 2015), p.4.

18. The underlying problems associated with Russia's fragmented design and building programme were highlighted in a report entitled 'Large Number of Classes in Russian Surface Fleet may cause Logistic Problems' carried by the *TASS* News Agency and reported on the *Navy Recognition* website on 26 May 2016.

19. The programme's current status was discussed by Nikolai Novichkov in 'Russia's Lada-class project suffers further delays', *Jane's Defence Weekly – 27 January 2016* (Coulsdon: IHS Jane's, 2016), p.14.

2.4A Fleet Review

Author:
Søren Nørby

THE ROYAL DANISH NAVY

FROM THE BALTIC TO THE HORN OF AFRICA – AND BACK AGAIN?

Founded in 1510, the Royal Danish Navy (*Søværnet*) is one of the oldest navies in Europe. Given Denmark is an island nation, the navy has always been an integral part of the country's defence. Before the British bombardment of the Danish capital Copenhagen and the subsequent seizure of the entire fleet in 1807, the Danish Navy was the sixth largest navy in the world. The navy thus builds on a long history; without the Danish Navy there would be no Denmark.

The Royal Danish Navy's support ship *Absalon* – essentially a highly-flexible frigate capable of performing a wide range of tasks – is typical of the small but sophisticated force of warships that have entered the fleet as the Danish Navy has shifted its focus since the end of the Cold War. *(Danish Defence)*

During the last twenty-five years or so, the Danish Navy has changed dramatically. When the Berlin Wall fell in 1989, the fleet consisted of approximately fifty warships. These covered a broad spectrum; from submarines to minelayers and from fast attack craft to frigates. Today the fleet has no submarines, no fast attack craft and no dedicated minelayers. Moreover, personnel numbers have been reduced from 5,600 in 1986 to c. 2,500 in 2015. However, a small number of significant new ships have been commissioned. Even though the fleet now consists of only five dedicated warships, they are the largest, most powerful and most versatile ships that the *Søværnet* has ever had at its disposal.

At the same time as the Danish Navy has gone through this transition, its focus has shifted from Danish waters and the Baltic towards the high seas. It has become a much-used tool for the Danish government and its new 'activist foreign policy'. The basic thinking behind this policy is that Denmark is inevitably affected by developments in the globalised world around it. Consequently, if the Danish government wants to help maintain peace and stability in Europe, it may be necessary to intervene in conflicts far beyond the immediate neighbourhood that was the focus of operations during the Cold War. This new approach has resulted in the navy becoming involved in war-fighting for the first time since the War of 1864.[1] Another aspect of this broad focus is the fact that a material percentage of the world's merchant shipping is comprised of Danish or Danish-owned ships.[2] As such, the spread

The Danish support ship *Esbern Snare* meets what was then the world's largest container ship, the Danish-owned *Mærsk Mc-Kinney Møller* of the Triple E-class, off the Horn of Africa in 2014. Denmark's global maritime and trading interests have supported a shift in the navy's focus away from home waters towards the high seas but its role as 'gatekeeper to the Baltic' means that it may need to counter a re-emerging threat from Russia. *(Danish Defence)*

The *Thetis* class patrol vessel *Væderen* pictured in Greenland in 2010. The Royal Danish Navy's constabulary responsibilities extend beyond Danish waters in Europe to the Faroe Islands and Greenland, which are both autonomous territories within the Danish Realm. *(J Nielsen)*

of maritime security problems such as piracy that have come more to the fore since the end of the Cold War can have a direct impact on the Danish economy, notwithstanding they can arise thousands of miles away.

In spite of these changes, Denmark cannot escape the importance of its geographical position in Europe as the 'gatekeeper to the Baltic'. Ships transiting to and from the Baltic must pass through the Danish Great Belt or the Sound between Denmark and Sweden.[3] Consequently, one of the *Søværnet*'s central roles has always been an ability to close the straits in case of an attack on the country or its allies. This was particularly crucial during the Cold War, as it was a major constraint to the actions of Soviet warships based in the Baltic. The significance of this role inevitably declined after the fall of the Berlin Wall but the strategic importance of Danish waters – and capabilities – remain important to Russian thinking. A clear indication was provided of this in March 2015 when the Russian ambassador to Denmark, Mikhail Vanin, wrote an 'op-ed' in the Danish newspaper *Jyllandsposten* making it clear that assigning Danish *Iver Huitfeldt* class frigates to participate in NATO's anti-ballistic missile defence (ABM) system would result in them becoming targets for Russian nuclear weapons. The renewed danger from Russia – made clear by its actions against Ukraine – has made it apparent that the shift from the Baltic to the high seas has come at a price. In simple terms, the Danish Navy is currently too small to handle a global role, whilst simultaneously countering a re-emerging threat from the east.[4]

An aerial view of the naval base at Frederikshavn, which was constructed in the 1960s with the help of NATO funding. One of two major Danish naval bases, it is home of the 1st Squadron, which is largely tasked with supporting ships operating in Danish territorial waters. *(Danish Defence)*

MISSION

The *Søværnet*'s overall mission is to protect Danish interests from the sea – with force if necessary. The navy is tasked with enforcing the Danish sovereignty in home waters whilst – at the same time – providing maritime rescue and environmental control when needed. The navy is assisted in these tasks by the Naval Home Guard (*Marinehjemmeværnet*), which operates thirty modern vessels.[5] With the navy's transition to much greater international engagement, the Naval Home Guard has taken over a large part of day-to-day maritime surveillance, rescue and environmental monitoring tasks in the Danish waters.

The Danish Navy is also tasked with patrolling the waters around the Faroe Islands and Greenland, which also form part of the Danish Realm. Ships here are tasked with maintaining a recognised surface and air picture of the Danish Kingdom's exclusive economic zone, including fishery protection, sovereignty enforcement and sea rescue over a vast area.

As already mentioned, the *Søværnet*'s principal units have become increasingly focused on international stabilisation missions in recent years.

ORGANISATION

Acting as the chief military adviser to the Danish Minister of Defence, the Danish Chief of Defence – currently General Peter Bartram – is the supreme military authority in the Danish Armed Forces and commands the Danish Defence Command. The Danish Naval Staff (*Marinestaben*) is an integrated part of the overall Defence Command organisation. In 2014, it was moved from Aarhus to the air base at Karup, where the Danish Army and Danish Air Force Staffs are also located.

Since 2013, the Chief of Naval Staff – the former Admiral Danish Fleet – has been Rear Admiral Frank Trojahn. He has a direct staff of approximately ninety personnel at his disposal. Amongst other commands reporting to him is a deployable command staff, the Danish Task Group, which was formed in 2001. Comprising a core of around twenty personnel, the unit has day-to-day responsibility for educating and training maritime forces as well as commanding deployed forces. It can be tasked with a wide range of assignments, ranging from crisis management, peacekeeping and peacemaking operations to the conduct of outright war. As a deployable capability, the group can lead international maritime operations at task group level either from platforms at sea or from ashore. This capacity has been exercised on frequent occasions, both during exercises and operationally (most notably during Combined Task Force 150 and 151 anti-terrorism and anti-piracy missions off the east coast of Africa).

The Danish Navy has two main naval bases. One is at Korsør on the Great Belt and one at Frederikshavn near the top of Jutland. Both were built during the 1960s with assistance from NATO funding and have fairly modern facilities. They are home to the so-called Operational Logistics Support Centres. In addition to operating base workshops, these can deploy dedicated expert teams globally in support of Danish warships, acting as forward base support units.

The navy's ships are organised into two Squadrons. The 1st Squadron in Frederikshavn is tasked with deploying ships and crews for the missions in Danish, Faroe Islands and Greenlandic waters. The 1st Squadron is also responsible for the navy's environmental ships, the royal yacht, the coastal rescue service and the hydrographic service. The 2nd Squadron, in Korsør, is focused on interna-

Table 2.4A.1: ROYAL DANISH NAVY – PRINCIPAL UNITS AS AT MID-2016

TYPE	CLASS	NO.	YEAR[1]	TONNAGE	DIMENSIONS	SPEED	CLASS MEMBERS
Fleet Escorts (5)							
Frigate – FFG	IVER HUITFELDT	3	2011	6,600 tons	139m x 20m x 5m	28+ knots	*Iver Huitfeldt* (F361), *Peter Willemoes* (F362), *Niels Juel* (F363)
Support Ship – FFG	ABSALON	2	2004	6,400 tons	137m x 20m x 6m	23 knots	*Absalon* (L16), *Esbern Snare* (L17)
Patrol Vessels (13)							
Patrol Frigate – FS	THETIS	4	1991	3,600 tons	113m x 14m x 6m	21 knots	*Thetis* (F357), *Triton* (F358), *Væderen* (F359), *Hvidbjørnen* (F360)
Ocean Patrol Vessel – OPV	KNUD RASMUSSEN	2	2008	2,100 tons	72m x 15m x 5m	17 knots	*Knud Rasmussen* (P570), *Ejnar Mikkelsen* (P571)
Ocean Patrol Vessel – OPV	AGDLEK[2]	1	1979	400 tons	31m x 8m x 3m	12 knots	*Tulugaq* (Y388)
Patrol Vessel – PB	DIANA	6	2007	c.250 tons	43m x 8m x 2m	18 knots	*Diana* (P520), *Freja* (P521), *Havfruen* (P522), *Najaden* (P523), *Nymfren* (P524), *Rota* (P525)
MCM Vessels (6)							
MCM Drone – MRD	MSF	4	1998	c. 125 tons	27m x 7m x 2m	12 knots	*MSF1, MSF2, MSF3, MSF4*
MCM – Drone – MRD	'HOLM'	2	2006	c. 150 tons	29m x 6m x 2m	12 knots	*Hirsholm* (MSD5), *Saltholm* (MSD6)
Auxiliaries (8)							
Survey Ship – AGSC	'HOLM'	2	2006	c. 150 tons	29m x 6m x 2m	12 knots	*Birkholm* (A541), *Fyrholm* (A542)
Training Vessel – AXL	'HOLM'	2	2006	c. 150 tons	29m x 6m x 2m	12 knots	*Ertholm* (A543), *Alholm* (A544)
Training Vessel – AXL	SVANEN	2	1960	c. 30 tons	19m x 4m x 2m	12 knots	*Svanen* (Y101), *Thyra* (Y102)
Diving Support Ship – AX	FLYVEFISKEN	1	1996	450 tons	54m x 9m x 3m	20 knots	*Søløven* (Y311)
Transport Ship – AKS	SLEIPNER	1	1986	500 tons	37m x 8m x 3m	11 knots	*Sleipner* (A559)

Other ships include four environmental protection vessels transferred to naval control in 1996 and the Royal Yacht *Dannebrog*. There are also a number of smaller service craft, personnel transports and tugs. The separate Naval Home Guard operates thirty coastal patrol vessels.

Notes:

1 Year refers to entry into service of oldest remaining vessel of a particular class.

2 Scheduled for replacement by the third *Knud Rasmussen* class OPV, *Lauge Koch*, during 2016.

tional operations, such as those off the Horn of Africa. Korsør is also home to the Danish Task Force organisation.

The navy's educational establishment – the Navy School – is headquartered at Frederikshavn and runs a number of subordinate, specialised centres covering areas such as marine safety, warfare, weapons and diving. The Naval Academy is now a subordinate unit – together with the Army Academy and Air Force Academy – of the Danish Defence Academy but remains situated at the old naval base of Holmen in Copenhagen. Holmen is also home to the naval band, the Centre for Diving Training and the Centre for Technical Training.

EQUIPMENT

The *Søværnet*'s principal vessels are listed in Table 2.4A.1. A central design feature of almost all major Danish warships is the StanFlex modular container system. Under this system, equipment such as weapons, sensors and cranes is placed in standardised containers that can be quickly slotted into pre-prepared shipboard positions depending on the requirements of a particular mission.

***Iver Huitfeldt* Class:** The three modern frigates of the *Iver Huitfeldt* class are the most powerful ships that the Danish Navy has ever had at its disposal. The ships are based at Naval Base Korsør. They are named after three Danish naval heroes.

The new frigates can perform a variety of tasks – from escort duties through to naval gunfire support for land operations – but are particularly potent in the air-defence role by virtue of the combination of their APAR and SMART radar systems and Mk 41 vertical-launch system (VLS) armed with Standard surface-to-air missiles. They are able to maintain a recognised air picture to c. 300km around the ship and could be adapted to participate in NATO's anti-ballistic missile defences in Europe. The Mk 41 VLS would also allow them to be armed with cruise missiles such as the United States' Tomahawk. However, the Danish parliament has yet to allocate funds for either acquisition. Other armament depends on which StanFlex modules are fitted but a typical loadout includes two 76mm Oto Melara guns, a 35mm close-in weapons system (CIWS), Harpoon and/or Sea Sparrow missiles and anti-submarine torpedo tubes. The ships also have hangar space for one embarked helicopter.[6]

Iver Huitfeldt has participated in anti-piracy operations off Africa whilst *Peter Willemoes* was part of the OPRECSYR force tasked with disposing of Syrian chemical weapons.

***Absalon* Class:** The two *Absalon* class vessels are designated as support vessels but are essentially flexible frigates, capable of performing a wide-variety of tasks. They preceded the *Iver Huitfeldt* class into service and the two classes share the same basic hull form.[7] However, in contrast to the *Iver Huitfeldt* class's sophisticated air-defence system, the central part of the *Absalon* class's design is a large – 900m² – flexi-deck that can be used to support a range of missions. For example, it can be used to transport armoured and other vehicles or even adapted to act as a holding facility, as has been the case during anti-piracy missions. Armament includes one 127mm Mk 45 gun, two 35mm CIWS and Sea Sparrow and/or Harpoon StanFlex containers. There is sufficient space to embark and sustain two helicopters.

Absalon and her sister, *Esbern Snare*, are named after two brothers who played a significant part in twelfth-century Danish history.

The frigate *Iver Huitfeldt* – seen here in 2012 – is the first of three new frigates with particularly powerful air defence capabilities. Along with the *Absalon* class, they form the core of the navy's war-fighting fleet. *(Torbjørn Kjosvold/ Norwegian Armed Forces)*

A close-up view of the Danish *Absalon* class support ship *Esbern Snare*. The later *Iver Huitfeldt* class air-defence frigates were derived from the design but incorporate a number of detailed differences by virtue of their different focus. *(Norwegian Armed Forces)*

***Thetis* Class:** Built principally for operations off Greenland and the Faroe Islands, the four vessels of the *Thetis* class are capable of operating in the Arctic all year round. They have been specially designed for operations in icy waters, featuring a double hull extending two metres above the waterline and ice-strengthening. Their principal armament is one 76mm gun. The class also have space for one organic helicopter. They are currently in the process of having their hangar rebuilt to fit the new Seahawk helicopters that are being acquired from Lockheed Martin's Sikorsky division in replacement for the veteran Lynx.

***Knud Rasmussen* Class:** Like the *Thetis* class, the two ships of the *Knud Rasmussen* class were designed for constabulary operations in the North Atlantic and around Greenland. They are also equipped with an Oto Melara 76mm gun, whilst two Stanflex container positions enables them to be fitted with additional weaponry, such as Sea Sparrow missiles or anti-submarine torpedoes. This allows them to be equipped for front-line missions in other waters should the need arise. The ships also have a helicopter landing deck aft but there is no hangar. Normal crew is just eighteen personnel, although up to forty-three people can be accommodated should the need arise. The ships are named after prominent Danish Arctic explorers.

A third ship of the class was ordered in 2013. Named *Lauge Koch*, she is expected to join the *Søværnet* before the end of 2016. She will replace the old ocean patrol cutter *Tulugaq*, which has served the navy since 1979.

***Diana* Class:** The six patrol vessels of the *Diana* class were built to patrol Danish territorial waters. *Diana* was commissioned in 2007, and the other five ships followed during 2008 and 2009. They replaced the eight aging *Barsø* class, which had served the navy since the 1970s. The *Diana* class vessels are c. 100 tons larger than the *Barsø* class and have twice their speed.

The ships are designed to perform a broad range of constabulary and support tasks. These include surveillance at sea, monitoring/reporting and maritime policing, peacekeeping, search and rescue (including treatment for decompression sickness), environmental pollution control and explosive ordnance disposal.

The *Diana* class are equipped with a single

The four members of the *Thetis* class were built primarily for constabulary operations off the Faroe Islands and Greenland and have strengthened hulls to allow deployment in icy waters. Their light armament reflects the lower threat associated with this primary tasking, although this could be upgraded if they needed to undertake frontline duties. This picture shows the second member of the class, *Triton*, on exercise with the Norwegian Armed Forces in 2010. *(Stian Lysberg Solum/Norwegian Armed Forces)*

StanFlex container position, but are not able to accommodate all types of containers due to configuration issues. The position is normally used for either a storage module or an anti-pollution module. They are lightly armed with one or two 12.7mm machine guns. The vessels are approved for operation in 10cm of ice.

Mine Countermeasures Assets: For the task of clearing mines, the *Søværnet*'s 2nd Squadron includes a unit named 'MCM Denmark'. Utilising a

The *Knud Rasmussen* class oceanic patrol vessels are used to supplement the *Thetis* class on constabulary duties around Greenland and in the North Atlantic. Two of the class have been commissioned to date and a third should be delivered before the end of 2016. *(Danish Defence)*

flexible, modular concept, this comprises a commanding officer and specialised personnel, containerised equipment – including a command module – that can be deployed ashore or on a seagoing unit, and specialised shipping. The last-mentioned includes four small (c. 125-ton) 'MSF' class boars fitted with minehunting sonar that can be remotely controlled from a mother unit and two slightly smaller 'Holm' class vessels that can be used either as control vessels or drones. The boats are not manned when acting as minehunting drones, so no personnel would be harmed if the boat triggers a mine. When a boat's sonar discovers a mine, a remotely-piloted mine disposal vehicle is dispatched carrying a small explosive charge, which is placed near the mine. The drone then returns to the mother unit and the charge is detonated and the mine destroyed. The MCM DK concept has been tested with great successes in multinational maritime exercises in and around the Baltic over the past few years.

The Danish Navy has adopted a modular mine countermeasures capability under the MCM Denmark organisation. A key component is a number of small, sonar-equipped vessels that can be remotely operated from a command module – located either on shore or afloat – during minehunting operations. This picture shows the drone *MSF3*. *(Conrad Waters)*

Mine countermeasures were previously in the hands of MCM-equipped members of the *Flyvefisken* or Standard Flex 300 class patrol vessels, which introduced the StanFlex modular concept into service. Fourteen of these vessels were commissioned into the navy from 1989 onwards but *Søløven* is now the only remaining vessel of the *Flyvefisken* class still in service. She was rebuilt in 2010–11 in order to serve as the navy's diving ship. During the modification, the ship's 76mm Oto Melara gun module was removed and the ship is currently only equipped with small arms. Other members of the class have been transferred to the Lithuanian and Portuguese navies.

The support ship *Esbern Snare* pictured operating with the roll-on/roll-off ferry *Ark Futura*. Under the Ark Project, Denmark and Germany have the ability to access five chartered civilian transport ships for military requirements. *(Danish Defence)*

Other Vessels: With over 100,000 ships passing through Danish waters in the course of a year, it should come as no surprise that control of marine pollution is also an important priority. In 1996 the task of fighting marine pollution in Danish waters was transferred from the Danish Ministry of Environmental Affairs to the Ministry of Defence. The navy therefore took control of the ministry's four pollution control vessels: two large 'Supply' class ships – *Gunnar Thorson* and *Gunnar Seidenfaden* – and two smaller 'Sea Truck' class boats, *Mette Miljø* and *Marie Miljø*. The ships are manned by the navy but painted in bright orange and yellow colours to signify their non-military role. Part of the 1st Squadron, one of each type is assigned to Naval Base Frederikshavn and the station at Copenhagen.

Two cutters, *Svanen* and *Thyra*, serve as training ships for the cadets from the Naval Academy. The cadets use them to learn basic skills such as navigation and seamanship, whilst also learning to work together as a crew. The ships also participate in various races. The Naval Academy also has two small training ships of the 'Holm' class at their disposal. A further two 'Holm' class vessels are used for inshore survey.

A Danish Navy Lynx helicopter embarked on *Absalon* pictured operating in the Mediterranean during the OPRECSYR mission aimed at disposing of Syria's chemical weapons stock-pile. The Lynx helicopter has now served the Danish Navy for over thirty years. *(Norwegian Armed Forces)*

The transport ship *Sleipner* (named after the Norse god Odin's horse) is the navy's only transport ship. Originally built for transporting torpedoes, it now serves as a workhorse for both the navy and army, transporting a variety of ammunition and other military stores.

A decision at the NATO summit in Prague in 2002 to increase the transport capacity of the alliance resulted in Denmark acquiring additional transport capacity. To reach this goal, in 2003 the navy founded the so called 'Ark Project', under which it chartered first one and, later, several more civilian transport ships from the shipping company DFDS. The Danish Navy has full rights to use these ships, but when not required by the navy, the ships sail in the normal DFDS network. In 2006 Germany also joined the Ark Project, which today has five chartered ships. These include the *Ark Futura*, which played a central role during OPRECSYR.

The Danish Royal Family is one of the few in Europe which still has a royal yacht. Manned by the navy the ship serves as a 'floating palace' and is widely used each summer when Her Majesty The Queen and the royal family visit a number of Danish ports. The ship also serves as the royal family's home during visits to the Faroe Islands and Greenland. Commissioned in 1932, *Dannebrog* has served the Danish Navy for longer than any other ship.

In addition to the vessels described, the navy owns a number of tugs and smaller support craft. The three icebreakers formally operated by the *Søværnet* were formally decommissioned at the start of 2013 following a government decision to cease government-operated icebreaking activities.

NAVAL AVIATION

The Danish Navy acquired use of its first helicopter in 1962. Eight Alouette III helicopters – initially operated as a flight of the air force's 722 Squadron – served mostly from ocean patrol vessels around Greenland and the Faroe Islands until 1982, where they were replaced with Westland Lynx. The Lynx has also served the Danish Navy with distinction; from the Arctic to the Horn of Africa. During much of this time, the helicopters served as part of the Naval Air – and subsequently Naval Helicopter – Service, which operated as an independent entity from 1977 to 2010. On 1st January 2011, however, the helicopters reverted to the Danish Air Force as 723 Squadron.

In 2012, Denmark decided to replace the Lynx with the American MH-60R Seahawk. The first three of nine Seahawks arrived in May 2016. The Seahawks are larger than the existing helicopters and

The Royal Danish Navy is replacing its current Lynx helicopters with the latest MH-60R version of the US Seahawk. A total of nine helicopters will be delivered by 2018. *(Lockheed Martin)*

Iver Huitfeldt (foreground) pictured exercising with the Royal Norwegian Navy's frigate *Fridtjof Nansen* in 2013. The Royal Danish Navy has acquired some powerful and flexible warships in recent years that provide potent capabilities for Denmark and its allies. For example, at least one of the *Iver Huitfeldt* class looks set to have its radar upgraded to act as a sensor as part of NATO's ballistic missile defence system. However, budget cutbacks have impacted operational effectiveness – only enough missiles have been acquired to equip one of the three air-defence frigates fully. *(T Dyhr)*

some ships require modifications to operate the new type. Due to budget constraints the new helicopters will not be equipped with anti-submarine sonar.

PERSONNEL

The Danish Navy employs about 2,500 men and women. It is a small force, and the navy is currently challenged by problems with retention. A high number of missions combined with better-paying jobs outside the military has resulted in a number of sailors and officers leaving the force. The navy – supported, of course, by the Armed Forces Personnel Department – is currently trying several measures to solve this problem but a number of jobs are still not filled.

Having spent many man-hours and resources in a low-intensity fight against pirates during the first decade of the current millennium, the *Søværnet* has, in recent years, taken a number of steps to strengthen its professionalism. For example, the frigates of the *Iver Huitfeldt* class have participated in – and passed – the British Navy's Flag Officer Sea Training (FOST) programme, whilst *Absalon* and *Esbern Snare* will also soon join this challenging training. At the same time, a number of crews have been trained at the German Navy's Damage Control Training Centre (DCTC) at Neustadt. Here they have learned important lessons about damage control at sea, and this has, amongst other things, led to a number of changes in ships' organisations.

FUTURE PROSPECTS

The Danish government allocated 20.8bn DKK (c. US$3.1bn) to the entire Danish defence budget in 2015. This amounted to c.1.2 percent of the Danish gross domestic product (GDP), which is quite far from the NATO goal of two percent GDP. Moreover, the defence budget was cut significantly in 2013, producing a number of challenges for the men and women in uniform. For example, the navy currently only has sufficient missiles to equip fully just one of the three *Iver Huitfeldt* class frigates.

Sometime during 2016, the Danish parliament is to make a decision about the type of fighter aircraft

The Royal Danish Navy pictured destroying a pirate mother ship during operations in the Indian Ocean in 2011. The navy's new ships are particularly well suited to long-range expeditionary and stabilisation missions but a smaller fleet will be faced with hard choices if more resources need to be devoted to ensuring security in home waters. *(Danish Defence)*

that will replace the ageing F-16, which has served Denmark since 1980. On 12 May 2016, the government announced it is recommending the purchase of twenty-seven F-35A Lockheed Martin Lightning II Joint Strike Fighters to meet the requirement. The new planes are expected to cost around 20bn DKK or c. US$3bn to acquire and have a total cost through a thirty-year lifecycle of more than 50bn DKK (US$7.5bn).[8] Given this funding will have to come from the existing defence budget, it is feared that this will mean there will be little money for new army and navy equipment in the years to come.

CONCLUSION

Whilst today's Royal Danish Navy is small, it operates some of the most potent vessels ever seen in the *Søværnet.* Moreover, these ships are particularly well-suited for the long-range, expeditionary type missions that have become increasingly common. The navy has certainly been challenged by budget cuts and by problems with personnel retention, but has still been capable of performing all the tasks set by the Danish government. However, hard choices and increasingly focused prioritisation have become the order of the day for The Chief of Naval Staff and it is questionable whether it will continue to be possible to do it all given ongoing financial constraints.

Moreover, during the last couple of years, there have been signs that the Baltic is again set to become a significant military hotspot given increasingly assertive Russian behaviour. If this indeed becomes the case, the *Søværnet* will find it difficult both to handle its increased commitment to international operations whilst undertaking the more local protection of Danish and allied interests in the Baltic. Ultimately, Denmark's politicians may be forced to choose between the two types of mission or find more money for naval resources.

The StanFlex 300 or *Flyvefisken* class patrol vessel *Støren* was one of fourteen ships delivered between 1989 and 1996. They introduced the StanFlex system of modularised weapons and other equipment that remains a central feature of Danish warship design. All but one of the class have now been withdrawn from Danish service as a result of the changed post-Cold War environment. However, more money may be needed to expand the fleet as the Baltic becomes more of a significant military hotspot. *(Conrad Waters)*

Notes

1. The War of 1864 – fought between Denmark on the one hand and Prussia and Austria on the other – ended disastrously for Denmark, with the loss of the mainland territories of Schleswig and Holstein. The Royal Danish Navy fought well in the war but was unable to have a material impact on the land campaign. The navy was subsequently involved in skirmishes with German occupying forces during the Second World War when the latter seized the fleet in August 1943.

2. The Danish A.P. Moller-Maersk Group controls the largest container shipping group in the world. Other significant shipping companies include D/S Norden and DFDS.

3. There are three channels connecting the Baltic to the North Sea through the Kattegat and Skagerrak: the Øresund between Denmark and Sweden, and the Great Belt and the Little Belt that run entirely through Danish waters. The last-mentioned has much more restricted navigation than the other two.

4. Some reference should also be made to the significance of Greenland, which – along with the Faroe Islands – is an autonomous territory within the Danish Realm. Another important strategic location during the Cold War, Greenland has assumed renewed significance due to man-made climate change. Notably, the melting of ice has made it increasingly possible to use the surrounding seas as transport links, as well as facilitating the potential exploitation of natural resources.

5. The Danish Home Guard, founded in 1949, is the fourth service of the Danish military and reports directly to the Danish Ministry of Defence. The Naval Home Guard (*Marinehjemmeværnet*) is one of its core constituents and has around 5,000 unpaid volunteers. Its patrol vessels are comprised of two basic classes – the MHV800 and MHV900 types. Both classes displace a little under 100 tons and are armed with light machine guns.

6. A detailed description of the design of the *Iver Huitfeldt* class was provided by Guy Toremans in '*Iver Huitfeldt* Class Frigates: Spearhead of the Royal Danish Navy', *Seaforth World Naval Review 2014* (Barnsley: Seaforth Publishing, 2014), pp.104–19.

7. The omission of the flexi-deck from the *Iver Huitfeldt* class means they have a lower profile. The aft section of the hull was also modified from that of the original *Absalon* class to accommodate the more powerful propulsion plant installed in the new frigates. In addition, the routing of uptakes had to be modified to avoid interference with the ship's air-defence radars.

8. This information is derived from *Executive Summary: Type Selection of Denmark's New Fighter Aircraft* (Copenhagen: Danish Ministry of Defence, 2016). Interestingly, the document suggests a procurement cost of DKK15.4bn for twenty-eight Joint Strike Fighters whilst the government's recommendation suggests an investment of DKK20bn for twenty-seven.

2.4B Fleet Review

Author:
Theodore Hughes-Riley

THE ROYAL NETHERLANDS NAVY

RECENT DEVELOPMENTS AND CURRENT STATUS

The Netherlands became the dominant world naval power in the seventeenth century, a status driven by their expansion into the 'New World' and the need to protect the maritime trade routes used for the lucrative spice trade. During this time the Dutch Empire acquired a number of colonial possessions across the globe, often wrestling them from the waning Portuguese Empire. Although Dutch naval power declined with the rise of other European fleets – particularly the British Royal Navy – the Netherlands remained a significant maritime empire. Even at the time of the Second World War the Kingdom of the Netherlands still retained substantial colonial holdings in both the Americas (Netherlands Antilles, Suriname) and the Pacific (the Dutch East Indies or *Nederlands-Indië*: the modern-day Republic of Indonesia).

The Royal Netherlands Navy's submarine *Walrus* – the first of a class of four boats – pictured departing Portsmouth, UK in March 2016. The Kingdom of the Netherlands is a wealthy trading nation that is dependent on global stability and freedom of the seas; maintenance of an effective naval service is therefore an important consideration. *(Conrad Waters)*

The Dutch colonial presence diminished with the formation of the Republic of Indonesia after a brutal war of independence following the end of the Second World War. Other nations emerged from the former colonies in the following decades. The remaining overseas territories – exclusively in the Caribbean – became integrated into the Kingdom of the Netherlands as constituent countries. Following the most recent constitutional reforms, the four countries of the Kingdom of the Netherlands currently comprise The Netherlands themselves, together with Curaçao, Aruba and Sint Maarten.[1]

The 21st-century Kingdom of the Netherlands is a wealthy nation. In nominal terms, it has the sixth highest gross domestic product (GDP) in Europe; around €650bn in 2015. This places it as the seventeenth wealthiest nation in the world. Much of this wealth is tied to the Kingdom's prominent role in world trade; imports and exports of c. €430bn in each direction position it amongst the world's ten largest trading countries. While many of the Netherlands' major trading partners are within

Europe, the kingdom still conducts a significant amount of trade with non-European nations, for example the United States of America and China. This makes maritime transport a critical component of its trading network. While the Netherlands has a number of large seaports, these are dwarfed by the Port of Rotterdam. This was the largest seaport in the world until 2004 and is still consistently ranked within the top ten busiest ports in the world. It is also the largest container port in Europe. The Netherlands' significant wealth, combined with the influence gained through membership of both the European Union and NATO, contribute to the Netherlands' status as one of the world's middle-ranking powers.

The interests of the Kingdom of the Netherlands are therefore intrinsically global, a fact that informs the nation's defence decisions, not least how to allocate a defence budget that currently stands at c. 1.2 percent of GDP. With the end of the Cold War, the main immediate defence concerns are global stability and combating illegal activities, as both can have a negative impact on their economy. The heavy reliance on trade – especially that conducted outside Europe – makes maintaining the freedom of the seas of paramount importance. This important role is primarily within the purview of the Royal Netherlands Navy (RNLN).

The Royal Netherlands Navy embarked on a major programme of construction to rebuild its war-ravaged fleet after the end of the Second World War. This image shows the lead ship of two *De Ruyter* class cruisers in home waters in September 1962. She remains in service to this day as the Peruvian Navy's *Almirante Grau*. *(Editor's collection)*

RECENT HISTORICAL BACKGROUND

The origins of the RNLN's current structure can be traced back to the aftermath of the Second World War and the need to reconstruct a war-ravaged fleet. As of 1947, the fleet comprised a single aircraft carrier, two cruisers, six destroyers, a frigate and supporting submarines and smaller vessels. With withdrawal from most of the Dutch East Indies, the fleet's main orientation shifted steadily towards Europe and NATO operations, for which two oceanic anti-submarine escorts groups led by a carrier or cruiser flagship would be provided in the event of conflict.[2] Further warships would be allocated to defensive operations in the North Sea. A large programme of domestic construction was put in place to provide the required modern shipping. This included the new *Holland* and *Friesland* class anti-submarine destroyers, as well as completion of two *De Ruyter* class cruisers ordered before the war to a modified design. Other ships were acquired from allied nations, most importantly the carrier *Karel Doorman* – the former British *Venerable*.

By the late 1960s, many of these ships were fast approaching retirement. Indeed, the Dutch abandoned carrier operations – partly due to cost – following a boiler room fire on *Karel Doorman* in 1968. Investment in maritime patrol aircraft provided a partial replacement. Significant efforts were also made to expand the surface anti-submarine force with modern frigates. By the mid-1970s a fleet of around twenty-five ships was envisaged, thereby providing scope for an additional escort group to be formed for deployment in the Channel and North Sea. The introduction of the *Tromp* class air-defence frigates – replacing the *De Ruyter* class cruisers – provided the fleet with a dedicated anti-air warfare platform to support the Atlantic escort groups.[3] This capability was subsequently reinforced by introduction of the two *Jacob van Heemskerck* class variants of the *Kortenaer* class frigates in 1986. In addition to these major surface combatants, the navy also maintained a force of six submarines and a considerable mine countermeasures force.

The RNLN's primary focus during this time was represented by its commitment to both NATO's Standing Naval Force Atlantic (STANAVFORLANT) escort group and its Standing Naval Force Channel (STANAVFORCHAN) mine countermeasures flotilla. Another important commitment was represented by the formation, in 1972, of the United Kingdom/Netherlands Amphibious Force (UK/NL AF), with the Netherlands essentially contributing a marine combat group to the British 3 Commando Brigade.[4] At this time the Netherlands Marine Corps had no dedicated amphibious ships and relied on merchant vessels, supply ships and allied amphibious shipping for transport. The RNLN was also required to carry out national standing tasks, including patrols of national territorial waters in Europe and the Caribbean.

Some reductions in planned force levels did occur during the later years of the Cold War. By the early 1990s, fewer than twenty frigates – albeit significantly more capable ships than their predecessors – were in service. The immediate post-Cold War era saw a much more significant re-orientation and restructuring, with both frigate and mine-countermeasure vessel numbers falling rapidly in an era where the rationale for significant numbers of such vessels was considered doubtful. The fleet of maritime patrol aircraft (MPAs) was also disbanded. A particularly damaging trend was a progressive reduction in planned force structures; for example the 2000 Defence White Paper's commitment to a force including fourteen frigates, twelve minehunters and modernised MPAs was quickly overturned. One small compensation was fulfilment of long-term ambitions to acquire specialised amphibious shipping: the amphibious transport dock *Rotterdam* commissioned in 1998 and a half-sister, *Johan de Witt*, followed some years later.

The most influential review of recent years has been that contained in the *Marinestudie* 2005 (Naval Study 2005). By this time the planned major surface combatant force structure had shrunk to four new *De Zeven Provinciën* 'LCF' type air-defence ships and six of an original fleet of eight *Karel Doorman* 'M' type frigates. The study envisaged the loss of a further four of the latter class, which were to be replaced by four new offshore patrol vessels – the *Holland* class. The new ships were designed to carry out intervention tasks in lower-threat areas and to carry out patrol duties in the Caribbean and North Sea, roles previously conducted by frigates.[5] The savings provided by this plan also allowed a replacement for replenishment capabilities through acquisition of the Joint Support Ship (JSS) *Karel Doorman* and implementation of previously planned improvements to minesweeping capabilities. More fundamentally, the plan completed the RNLN's transformation from a navy focused on Cold War anti-submarine operations to a structure orientated towards constabulary and expeditionary missions.

Although the onset of global recession and the Eurozone's subsequent economic problems threatened to derail major parts of the plan – at different times two of the four *Holland* class and *Karel Doorman* have been earmarked for sale – the Naval Study 2005 has broadly stood the test of time. However, economies announced in 2011 did see the implementation of some further force reductions. These included a cut in the mine-countermeasures force to just six ships and the withdrawal of the replenishment ship *Amsterdam*, the latter now sold to Peru.

CURRENT DEFENCE PRIORITIES AND COLLABORATIONS

The Kingdom of the Netherlands has three defence priorities, initially set out in the 2000 Defence White Paper:

- Protecting the integrity of national and Allied territory, including the Netherlands Antilles and Aruba.
- Promoting stability and the international rule of law.
- Supporting civil authorities in upholding the law, providing disaster and humanitarian relief, both nationally and internationally.[6]

The White Paper described the ambitions of the government to fulfil these tasks, making reference to defence within the NATO context, limited duration participation in peace-enforcement operations at brigade level or equivalent (e.g. a maritime task group), long-term participation in peacekeeping operations at a lower level, the protection of sovereign territory, and supporting civil tasks. The white paper made explicit mention of protecting the Netherlands territories in the Caribbean and supporting coastguard duties to combat the proliferation of drugs.

In practice, overall force composition and the strength of predicted contributions to operations has declined substantially in the sixteen years since the White Paper, with significant defence cuts impacting all three services. These reductions have been widely criticised by military commentators, as they reduce the nation's ability to address both its own security concerns as well as its commitment to NATO.[7] From a naval perspective the significantly smaller surface fleet has already limited the RNLN's ability to conduct core tasks. As recently as 2014 – when a task force centred on the Russian carrier *Admiral Kuznetsov* travelled through the English Channel towards the North Sea – the RNLN was unable to provide a vessel to shadow the force; an event widely reported on by the media.

Nevertheless, the RNLN makes an important contribution to the Dutch government's defence priorities, not least the policing of the Kingdom's exclusive economic zone (EEZ) in both the North Sea and the Caribbean. Additionally, the significant RNLN contribution to NATO has not diminished in the aftermath of the Cold War. NATO's increasing interest in 'out of area' operations – in particular, in protecting shipping lanes – synergises with the interests of the Dutch government. For example, Dutch frigates are regular participants in the two NATO Standing Maritime Groups that contributed to Operation 'Ocean Shield' in the Indian Ocean between 2009 and 2014. The RNLN also provides a regular contribution to Standing NATO Mine Countermeasures Group 1. The Kingdom of the Netherlands also has strong ties with NATO members outside of the context of the organisation that are crucial to its maritime strategy.

The 1970s saw the Netherlands renew its fleet with modern surface escorts, including specialised air-defence frigates. This image shows the new *De Ruyter* – the second of two *Tromp* class frigates – towards the end of her operational career in 2001. The class was replaced – along with the pair of *Jacob van Heemskerck* type frigates – with the four new 'LCF' air-defence ships. *(Conrad Waters)*

More broadly, it is an active participant in the European Union-led EUNAVFOR, which conducts anti-piracy operations off of the Horn of Africa and Western Indian Ocean under the banner of Operation 'Atlanta'. Since 2015 the RNLN's aspirations for contributing to global operations has been either a single task force of five vessels, two ships deployed separately, or the deployment of specialist capabilities such as submarines.

The 2005 Netherlands Defence Doctrine marks out the United States, United Kingdom and Belgium as principal allies in the maritime domain. Collaboration with US armed forces is particularly strong in the Caribbean; notably an agreement between the Dutch armed forces and United States to combat the drug trade in the region was first signed in 1999. The Kingdom of the Netherlands is an ongoing participant in the United States Joint Interagency Task Force South (JIATFS), which tackles the narcotics trade – as well as other illicit trafficking such as that in humans, weapons and money – from South America to countries in Europe and North America. The Dutch Commander in Chief in the Caribbean is a subordinate JIATFS commander in charge of Task Group 4.4. Meanwhile, the strongest obvious link with the United Kingdom is through the UK/NL AF. Over the decades since the force was conceived the British Royal Marines and Dutch Marine Corps (*Korps Mariniers*) have pursued integration in a number of areas including logistics and materials. In addition, the Kingdom of the Netherlands has agreed to join the high-readiness UK Joint Expeditionary Force, which is due to become operational by 2018.

However, the nation with which the RNLN has the greatest synergy is Belgium and its Belgian Naval Component. This co-operation dates from the post-war Belgium-Netherlands Co-operation (BENESAM) Accord first signed in 1948, which provided for the two navies to operate under a single command in time of war. Collaboration has steadily strengthened over the years, culminating in the establishment of a single command under the leadership of Admiral Benelux in 1996.[8] The position of Admiral Benelux is held by the Commander of the Royal Netherlands Navy (currently Marine Corps Lieutenant General Rob Verkerk) and the position of Deputy Admiral Benelux by the Commandant of the Naval Component (a Belgian officer with the rank of Flotilla Admiral), with the staff located at the main Dutch naval base in Den Helder. This close relationship has allowed for the efficient use of materials and training. Both navies currently operate *Karel Doorman* class frigates and 'Tripartite' class mine-countermeasures vessels and are likely to collaborate on replacement programmes.

An important priority for the Netherlands' armed forces is the protection of national territory, including those parts of the Kingdom located in the Caribbean. An ongoing naval and coast guard presence is maintained in the region, not least the Curaçao-based support ship *Pelikaan*. She is pictured here taking part in relief operations following the Haiti earthquake of 2010. *(Royal Netherlands Navy)*

A new international partner in the naval arena is likely to be Germany following signature of a letter of intent in February 2016 that will see shared German use of the JSS *Karel Doorman* and integration of the German Navy's force-protection orientated sea battalion into the RNLN's command structure.

MANNING AND ORGANISATION

The Netherlands' armed forces are now entirely professional (i.e. non-conscript), with the RNLN having c. 9,250 full-time personnel and c. 850 reservists.

Reporting to the Commander of the Armed Forces (Chief of Defence Staff), the Commander of the Royal Netherlands Navy is based at the *Nieuwe Haven* (New Port) in Den Helder. Also located there are the rest of the Admiralty, including the Deputy Commander and the three main Directorates of Operations, Materiel & Support, and Personnel & Administration. The Admiralty is also responsible for the separate Netherlands and Dutch Caribbean coast guards, the latter being headed by the Commander of the Navy in the Caribbean.

With the exception of the support ship *Pelikaan*, which is permanently assigned to the Caribbean, Den Helder is home to the entire Dutch fleet. It is also the location of a number of important support operations, such as the Diving Medical Centre. There are, however, a number of secondary naval facilities on the Dutch mainland. In the Caribbean, the main centre of operations is at the Parera Naval Base in Curaçao.

In addition to the operational fleet (see further below), significant RNLN and associated operational assets include:

Marine Corps: The Marine Corps falls under the RNLN's Operations Directorate; the Commandant of the Marines also being Director of Operations.

Troops from the Dutch Marine Corps pictured operating in the training areas of the island of Texel in April 2014. The Netherlands can draw on a Marine Corps totalling some 2,300 personnel. *(Royal Netherlands Navy)*

Since 2008, the RNLN's air component has been absorbed into a tri-service helicopter command. 860 Squadron provides flights for shipboard operations: this picture shows one of its new NF-90 helicopters in 2012. *(Royal Netherlands Navy)*

Currently totalling 2,300 personnel, it is focused around two Marine Combat Groups (726 personnel each). Although the Marine Corps specialises in maritime warfare, the combat groups have participated in a number of land-based deployments, including to Iraq and Afghanistan, in a light infantry role in recent years. Other formations include the Netherlands Maritime Special Operations Forces (which focus on counter-terrorism, underwater operations and mountain warfare), the Surface Assault & Training Group (responsible for landing and fast interception craft), the Sea-Based Support Group and an independent squadron based in Aruba. Major facilities are currently located at Rotterdam (basic training), Doorn near Utrecht, Den Helder and Texel.

Netherlands Maritime Force: A deployable headquarters consisting of around sixty personnel (with the possibility of increasing staffing to over 150 should the need arise), Netherlands Maritime Force (NLMARFOR) is also used to help train ships and marine units prior to operational deployment. Its multinational staff includes US, Belgian, French and UK personnel as well as a range of RNLN specialists. NLMARFOR has no units permanently assigned to it but is designed to work flexibly with a range of national and allied units.

Defence Helicopter Command: The RNLN's former discrete air component was absorbed into the tri-service defence helicopter command in 2008. The maritime helicopters have remained at De Kooy Airfield (near Den Helder), a joint civilian and military airport. De Kooy is home to four squadrons, two of which are flying units: 7 Squadron – a training and standards unit – and 860 Squadron which provides flights for seaborne operations and search and rescue. Both squadrons operate the NH-90 helicopter. Of the twenty purchased, twelve are NATO Frigate Helicopter (NFH) variants, and eight are TTH transport variants for use by the Marine Corps. The Defence Helicopter Command structure also facilitates deployment of other helicopters such as the Chinook-47D (298 Squadron) or the AH-64D Apache (301 Squadron) on appropriate naval vessels.

THE FLEET

The current composition of the RNLN is shown in Table 2.4B.1, with details of the major units discussed below.

***De Zeven Provinciën* Class Frigates:** The *De Zeven Provinciën* class frigates provide the war-fighting backbone of the Dutch fleet. While sporting the title of a 'frigate', the class members are c. 6,000-ton anti-air warfare and command vessels that many other navies would call a destroyer.[9] Their command facilities are suitable for hosting the staff of a maritime task group commander. The design originates from a joint project with Spain and Germany that also gave rise to the *Álvaro de Bazán* (F-100) and *Sachsen* (F-124) classes and the Dutch ships share much in common with the latter design.

In their primary air-defence role the class makes use of a Thales Nederland SMART-L long-range surveillance radar in conjunction with a Thales Nederland multifunction APAR (Active Phased-Array Radar). The two radars – operating in different frequencies – complement each other. The impressive SMART-L's ability to form a comprehensive air picture encompassing up to a thousand objects at distances out to c. 500km essentially supports APAR's more precise tracking and missile guidance capabilities. The radars operate in conjunction with the vessels' forty-cell Mk 41 vertical-launch system, which has a theoretical load-out of thirty-two Standard SM-2 area air-defence missiles and a further thirty-two quad-packed Evolved Sea Sparrow missiles (ESSM) for shorter-range protection. The ships also carry one or two indigenous Goalkeeper close-in weapons systems (CIWS) for last-ditch defence as well as a balanced anti-surface and anti-submarine armament.

In addition to its conventional air-defence role, the SMART-L radar has the proven ability to track a ballistic missile at extended ranges. Extensive work is being undertaken on upgrading the class to acquire a ballistic missile tracking capability, which is expected to be introduced by late 2017. This will form the Netherlands' contribution to NATO's ballistic missile defence capability. At the present time, however, Dutch ministers have only committed to the implementation of the necessary radar upgrades to allow ballistic missiles to be tracked at long ranges – to date no commitment to purchase the Standard SM-3 missiles required to destroy a ballistic weapon has been made.

Despite their high-end credentials, the class has seen extensive use on constabulary duties, including participations in both Operation 'Ocean Shield' and Operation 'Atlanta' in the Indian Ocean. The former deployments have often taken place as part of the

The core of the RNLN's surface forces is comprised of the four anti-air warfare and command frigates – destroyers in all but name – that were delivered between 2000 and 2003. This picture shows *Evertsen* – the last member of the class to be delivered – in Scottish waters during a NATO exercise in April 2013. *(Royal Netherlands Navy)*

The lead *De Zeven Provinciën* class 'LCF' anti-air warfare and command frigate pictured in July 2009. Work is being undertaken to upgrade the SMART-L radar positioned on the short mainmast to provide a ballistic missile tracking capability. *(US Navy)*

Table 2.4B.1: ROYAL NETHERLANDS NAVY – PRINCIPAL UNITS AS AT MID 2016

TYPE	CLASS	NO.	YEAR[1]	TONNAGE	DIMENSIONS	SPEED	CLASS MEMBERS
Fleet Escorts (6)							
Frigate – FFG	DE ZEVEN PROVINCIËN	4	2002	6,100 tons	144m x 19m x 5m	30 knots	*De Zeven Provinciën* (F802), *Tromp* (F803), *De Ruyter* (F804), *Evertsen* (F805)
Frigate – FFG	KAREL DOORMAN	2[2]	1993	3,300 tons	122m x 14m x 5m	29 knots	*Van Amstel* (F831), *Van Speijk* (F828)
Patrol Vessels (4)							
Oceanic Patrol Vessels – OPV	HOLLAND	4	2012	3,800 tons	108m x 16m x 5m	22 knots	*Holland* (P840), *Zeeland* (P841), *Friesland* (P842), *Groningen* (P843)
Amphibious Ships (2)							
Amphibious Transport Dock – LPD	ROTTERDAM	1	1998	12,800 tons	166m x 27m x 6m	21 knots	*Rotterdam* (L800)
Amphibious Transport Dock – LPD	JOHAN DE WITT	1	2007	15,500 tons	176m x 29m x 6m	19 knots	*Johan de Witt* (L801)
There are also 5 LCU, 12 LCVP and large numbers of smaller raiding craft							
Joint Support Ships (1)							
Joint Support Ship – JSS	KAREL DOORMAN	1	2015	28,000 tons	205m x 30m x 8m	18 knots	*Karel Doorman* (A833)
Mine Countermeasures Vessels (6)							
Mine Countermeasures – MCMV	ALKMAAR	6[3]	1985	650 tons	52m x 9m x 3m	13 knots	*Makkum* (M857), *Schiedam* (M860), *Urk* (M861), *Zierikzee* (M862), *Vlaardingen* (M863), *Willemstad* (M864)
Auxiliaries (6)							
Survey Ships – AGS	SNELLIUS	2	2003	1,900 tons	75m x 13m x 4m	12 knots	*Snellius* (A802), *Luymes* (A803)
Logistic Support Ship – AP	PELIKAAN	1	2006	1,200 tons	65m x 13m x 3m	15 knots	*Pelikaan* (A804)
Submarine Support Ship – ASL	MERCUUR	1	1987	1,400 tons	64m x 12m x 4m	14 knots	*Mercuur* (A900)
Training Vessel – AXL	VAN KINSBERGEN	1	1999	650 tons	41m x 9m x 3m	13 knots	*Van Kinsbergen* (A902)
Training Vessel – AXL	URANIA	1	1938	75 tons	27m x 6m x 3m	12 knots	*Urania* (Y8050)
There are also 5 diving support vessels and various tugs and other harbour craft							
Submarines (4)							
Patrol Submarine – SSK	WALRUS	4	1992	2,800 tons	68m x 8m x 7m	20 knots	*Walrus* (S802), *Zeeleeuw* (S803), *Dolfijn* (S808), *Bruinvis* (S810)

Notes:

1 Year refers to entry into service of oldest remaining vessel of a particular class.

2 Survivors of a class of 8.

3 Survivors of a class of 15.

RNLN's contribution to the NATO standing maritime groups, for which the class's command facilities are particularly useful. The class has also seen service on patrol duties in the Caribbean.

The four members of the class were commissioned between 2002 and 2005 and have an expected service life of at least thirty years. As such, the vessels are not expected to be replaced until the 2030s at the earliest.

***Karel Doorman* Class Frigates:** The *Karel Doorman* class (or 'M' – multi-purpose – class) frigates are general-purpose vessels that were originally conceived to meet a requirement for a frigate focused on North Sea operations but eventually grew into a much more capable design. Eight of these vessels were originally built for and operated by the RNLN, with early sales due to defence cuts seeing only two remaining with the service. Two sister-ships now make up the major surface combatant force of the Belgian Naval Component; two other pairs serve with Chile and Portugal.

Despite being a 'general-purpose' vessel these ships were designed during the Cold War and – given the RNLN's focus at the time – were particularly well-equipped for anti-submarine warfare. Upon commissioning, they were equipped with a powerful underwater sensor suite comprising both an active hull-mounted sonar and a passive towed array (the Thompson Sintra Anaconda DSBV 61). This coupled with its two twin torpedo tubes (US Mk 32 tubes armed with the Honeywell Mk 46 Mod 5 torpedo) and – particularly – an embarked ASW helicopter (initially the Westland Lynx) provided the vessels with an effective arsenal against sub-surface threats.

The two vessels of the *Karel Doorman* class remain the RNLN's most effective anti-submarine units, albeit its current sea-control helicopters – the NFH

A picture of the *Karel Doorman* or 'M' class frigate *Van Ness* in the Caribbean in 2007. There were originally eight ships of the class in Dutch service but they have steadily been transferred to other navies – *Van Ness* is now in service with the Portuguese fleet as *Bartolomeu Dias*. *(US Navy)*

The RNLN's amphibious forces have expanded in recent years; this image shows the amphibious transport dock *Rotterdam* during a visit to Takoradi, Ghana in September 2013. *(US Navy)*

variant of the NH-90 – are very effective anti-submarine platforms in their own right. Moreover, the RNLN has shown a commitment to maintaining the remaining ships' high-end anti-submarine capability through a programme with Ultra Electronics to develop and deploy an advanced Multi-static Active Passive Sonar system that operates at very low frequencies. After extensive trials with a prototype, the programme bore fruit when *Van Amstel* received the first production model of the sonar in 2014 as part of its mid-life upgrade. Also included in the refit were modifications to aviation facilities to allow for the operation of the NH-90, as well as the replacement and upgrade of various aging sensor systems to allow the vessel to remain credible until the end of an expected lifetime in 2020s.

***Holland* Class Offshore Patrol Vessels:** As discussed earlier in this chapter, the *Holland* class offshore patrol vessels were procured as a replacement for four of the *Karel Doorman* class frigates. Their 3,750-ton displacement is greater than that of the *Karel Doorman* class – partly due to the adoption of commercial construction standards and the use of heavier steels – but they lack the weapons fit and towed sonar array of the front-line frigates. Despite this, the vessels are very well suited for their intended constabulary role, with their Thales integrated mast (I-mast) providing powerful surveillance capabilities.

Amphibious Ships: The RNLN's amphibious support ships comprise the LPD-type amphibious transport docks *Rotterdam* and *Johan de Witt*. Whilst both ships are derived from a collaborative effort between the RNLN and Spanish Navy, they were laid down seven years apart and there are significant differences between them. For example, *Rotterdam* has direct diesel-drive whilst her larger half-sister incorporates podded propulsors. Both vessels were designed to act as sea-based command platforms, and possess the most comprehensive facilities of this type of any vessels within the RNLN. However, significant improvements were incorporated in *Johan de Witt*'s command suite and she is capable of commanding larger and more sophisticated operations.

In their main role as amphibious ships, the two vessels can transport c. 550-600 troops, roughly equivalent to a Marine Combat Group. Interestingly, *Rotterdam* has a slightly higher capacity in spite of being the smaller ship. Troops can be disembarked by helicopter or landing craft. *Rotterdam* can accommodate up to two landing craft utility (LCU) and three smaller landing craft vehicle personnel (LCVP) in a well deck. *Johan de Witt* has a slightly different arrangement by which two LCUs are housed in the well deck and four LCVPs launched by davits. Both ships can embark up to six NH-90-sized helicopters in a spacious hangar.

The vessels are armed with two Goalkeeper CIWS for self-defence against missile attack and a range of machine guns to protect against asymmetric threats. The ships have both been deployed on counter-piracy operations, in addition to use on counter narcotics missions and for disaster relief. Both ships are equipped with Role 2 hospitals.[10]

***Karel Doorman* Class Joint Support Ship:** The name of this ship, previously used by the RNLN's only two aircraft carriers, was perhaps chosen as a

The Dutch amphibious assault ship *Johan de Witt* pictured off Freetown in Sierra Leone in October 2009. She is a larger half-sister of the *Rotterdam*, with improvements that include improved command and control facilities and a strengthened flight deck. *(US Navy)*

The new JSS *Karel Doorman* is the largest ship in the Royal Netherlands Navy. She is intended to combine fleet replenishment, transportation and amphibious support facilities in a single hull. *(Derek Fox)*

statement of intent. At 28,000 tons, the new *Karel Doorman* is the largest vessel ever to be operated by the RNLN. Whilst essentially a replacement for the replenishment oilers *Amsterdam* and *Zuiderkruis*, she is styled as a Joint Support Ship (JSS). This better highlights her multi-role design and potential role as a transport, support ship and 'sea-base' for amphibious operations.

As such, *Karel Doorman* has significant capability enhancements over her predecessors to support her amphibious role. These improvements include the ship's large hangar space – allowing for up to six NH-90s' – and positions for two davit-launched LCVPs and two RHIB interceptors. The vessel also has a roll-on/roll-off capability for vehicle transfer and a berthing point – but not a full well deck – for a LCU. There is a substantial vehicle deck for the transport of tracked or wheeled vehicles or containers. As might be expected from a ship replacing replenishment tankers, *Karel Doorman* is also fully capable of replenishment-at-sea (RAS), incorporating two RAS masts and almost 9,000 tons of fuel capacity for this purpose. Sensors are housed in an integrated mast identical to that found in the *Holland* class. As for the amphibious ships, armament is limited to that needed for close-in self-defence.

To reduce costs associated with manpower the vessel makes use of 'crew modules' where, in addition to the core crew required to sail, additional crew – such as medical teams or an aviation detachment – can be brought aboard to activate facilities depending on mission. Equipment includes a Role 3 hospital.[10]

Despite the capabilities that this vessel brings to the fleet, plans to bring the vessel into service were almost halted in 2013 due to defence cuts that were announced whilst *Karel Doorman* was still fitting out. Although this decision was quickly revised, commissioning progressed on the basis that the ship would only be crewed to carry out the replenishment role. Use of her additional features would be dependent on collaboration with other allied nations. As previously mentioned, the recent signature of a letter of intent with Germany for shared use of the ship has brought this ambition closer to fruition.

Small Surface Vessels: The RNLN's front-line warships are supported by the usual force of second-line and support vessels. The most numerous – and important – of these vessels are the remaining six 'Tripartite' or *Alkmaar* class mine countermeasures

The Dutch mine countermeasures vessel *Willemstad* operating with other NATO units as part of Standing NATO Mine-Countermeasures Group 1 (SMMCG1) in the Baltic in May 2015. *(US Navy)*

vessels, which were originally built as part of a joint programme with Belgium and France. These remaining vessels have received extensive upgrades to their sonar and mine-disposal systems and are expected to remain in service for another decade or more. There are also four *Cerberus* class diving support vessels that support the mine countermeasures service, as well as a former army support vessel used for diver training.

Support vessels include the submarine support ship and torpedo tender *Mercuur* and the logistic support vessel *Pelikaan*. The latter, commissioned in 2006, is based at the Parera Naval Base in Curaçao and provides sealift services for marine and army units in the Caribbean. She was used to provide relief supplies after the 2010 Haiti earthquake. There are also two *Snellius* class survey ships, two training ships (one of which is for sail training) and a collection of tugs and harbour craft.

Submarines: The RNLN's submarine service comprises the four 2,800-ton (submerged) *Walrus* class boats, the first of which was commissioned in 1990. These diesel-electric submarines are powered by three SMET-Pielstick diesel engines connected to alternators that provide electricity to a single Holec motor. They are widely regarded as being very silent boats. This combined with a capability for conducting blue-water operations has put them in high demand with NATO. The submarines are armed with four 533mm torpedo tubes, with a normal load-out of up to twenty Mk 48 torpedoes providing a powerful underwater punch. The boats are also equipped 'for but not with' the Sub-Harpoon anti-surface missile, this variant of the missile never actually having been purchased by the Dutch Government.

Since the Cold War the submarines have been involved in a variety of roles. While many have been classified intelligence gathering operations, they have also seen publicised deployments to the Caribbean, Mediterranean and Indian Ocean as part of operations against piracy, terrorism and organised crime.

In 2013 a long-awaited programme to undertake a life-extension to the four submarines got underway, extending their lives until around 2025. In addition to anti-corrosion work, sonar and other surveillance

Mine countermeasures has been an important role for the Royal Netherlands Navy since the time of the Cold War. Although the fleet of mine countermeasures vessels has shrunk to just six ships, all are well-equipped. This image shows *Zierikzee* in 2011. The ships are one of a number of classes that will need replacement over the next decade. *(Conrad Waters)*

The RNLN can draw on a range of support vessels to assist its front-line operations. *Mercuur*, pictured here, is a specialist submarine support ship and torpedo tender. *(Conrad Waters)*

systems will be enhanced and integrated with a modernised combat management system. This will result in an improvement to the submarines intelligence gathering facilities, making them more suitable for the roles in which they are now typically employed. Upgrades to the fleet are well underway and expected to be completed by 2019.

FUTURE EQUIPMENT REQUIREMENTS

The RNLN will face the need to replace a number of major vessels in the mid-2020s, viz. the two remaining *Karel Doorman* class frigates, the *Walrus* class submarines, and the *Alkmaar* class minehunters. In addition, background information to the 2015 budget also made reference to a requirement to replace the amphibious transport docks in due course. At present, it appears plans for the frigate and submarine replacements are the most advanced, albeit budget constraints appear to be pushing back actual procurement decisions.

Some details have been revealed about the proposed *Karel Doorman* class frigate replacement (dubbed the Future Surface Combatant), including a conceptual model released by the Defence Materiel Organisation (DMO). Given the anti-submarine orientation of the vessels that they will replace and the significant funds invested in the new Multi-static Active Passive Sonar, it is probable that the system will be ported over from the *Karel Doorman* class to provide the new ships with an impressive submarine locating capacity. Belgium, also current operators of *Karel Doorman* class frigates, would be an obvious collaborative partner on the replacement project, particularly as replacement of the ships was confirmed in December 2015 as part of the Strategic Defence Plan for 2030. There has to be some doubt as to where the new ships will ultimately be fabricated given the use of the Dutch Damen group's facility at Galati in Romania for much recent new construction.

The *Walrus* class submarine replacement is also only in its early stages. This project has raised the possibility of partnership with a number of potential international collaborators and this seems a likely way forward. Currently it seems probable that Sweden will be the preferred partner given Saab Kockums (the manufacturers of Sweden's submarines) and Damen (the principal Dutch warship manufacturer through its ownership of the former Royal Schelde yard at Vlissingen) signed a deal to work together towards the 'potential' *Walrus* class replacement in 2015. This may lead to the Dutch replacement sharing much in common with the future Swedish Navy A26 submarine. Given that the last submarines to be built in the Netherlands were completed in the early 1990s, it is quite possible that the new boats will be built abroad.

CONCLUSION

Ongoing cuts to the Dutch defence budget have had a dramatic effect on the size and effectiveness of the RNLN, as they have for the other branches of the armed services. Even if the overall budgetary position is now more stable, operational requirements may mean the navy does not receive the funding it needs.

The large replacement programme needed to replace life-expired equipment in the current fleet may impose particular difficulties. Replacements for the many ships due to retire in the mid-2020s have yet to be ordered and planned dates have slipped, most recently in the 2016 defence budget. There is a risk that some programmes may not be realised.

In spite of these issues, the modern RNLN is an effective green-water navy. Whilst the fleet is smaller than it was in the Cold War, it fields a fleet of highly advanced vessels that, taken together, combine an ability to undertake a wide range of missions with some important niche capabilities. For example, the fleet has particular strengths with respect to both amphibious operations and high-end constabulary missions. All-in-all, this balance of capabilities makes the RNLN a vital NATO constituent, a position that it is likely to hold for the foreseeable future.

A picture of the modernised 'M' class frigate *Van Amstel* firing an Evolved Sea Sparrow Missile (ESSM). The two remaining ships of the class require replacement over the course of the next decade. However, funding is tight and the programme has already been delayed. *(Royal Netherlands Navy)*

Notes

1. The three remaining Dutch islands in the Caribbean – Bonaire, Saba and Sint Eustatius – became special municipalities of the Netherlands in 2010.

2. This section can only provide a brief overview of a relatively complex subject. For more details of development in the earlier and mid Cold War period, see John Jordan's chapter on 'The Netherlands' in Robert Gardiner (ed), *Conway's All the World's Fighting Ships 1947-82 – Part 1: The Western Powers* (London: Conway Maritime Press, 1983), pp.78–91.

3. One of the *De Ruyter* class cruisers that the *Tromp* class replaced had been refitted with the RIM-2 Terrier surface-to-air missile system in the 1960s. However, the *Tromp* class were the first vessels designed as air-defence escorts from the outset.

4. Hans Boersma has produced a website on the order of battle of the Netherlands Armed Forces as they stood in 1985. The section on the Marine Corps provides interesting insight into the operational role of the Marine Corps within NATO in that year and is available at: http://www.orbat85.nl/order-of-battle/royal-navy/marine-corps/kmarns.html

5. The editor's 'Holland Class OPVs: The "Rolls-Royce" of Offshore Patrol Vessels', *Seaforth World Navy Review 2016* (Barnsley: Seaforth Publishing, 2015), pp.89–103 provides a more in-depth discussion of the origins of the *Holland* class OPVs.

6. See *Defence White Paper 2000* (The Hague: Ministry of Defence, 1999), p.53. This section has been informed significantly by the wealth of English-language documents made available by the Government of the Netherlands, notably 2005 *Netherlands Defence Doctrine* (The Hague: Netherlands Defence Staff, 2005). See: https://www.government.nl/documents/leaflets/2010/01/12/netherlands-defence-doctrine.

7. An informative article on the decline in Dutch defence spending was published in 2013 by Marcial Hernandez of the American Enterprise Institute under the title: *Dutch hard power: Choosing to decline*. A link to the article is currently available at: http://www.aei.org/publication/dutch-hard-power-choosing-decline/?utm_source=twitter&utm_medium=web&utm_campaign=040313

8. The position of Admiral Benelux was first established in 1975, but only as a purely wartime role.

9. The ships are often known as the LCF class, an abbreviation of *Luchtverdedigings- en commando fregatten*, the Dutch for 'air-defence and command frigates'.

10. Under standard NATO terminology, there are four tiers ('Roles') of medical support, defined on the basis of capabilities and resources. Role 4 is the most extensive tier of medical support and Role 1 the least. Role 3 medical support is usually the most extensive found in a theatre of operations (in a navy, typically on a major amphibious ship or a hospital ship) and will include specialist surgical capabilities. Role 2 denotes a lower level of capability, although beyond that of the basic first aid, lifesaving resuscitation and triage available under Role 1 medical support.

115

3.1 SIGNIFICANT SHIPS

Author:
Tomohiko Tada

AKIZUKI CLASS DESTROYERS

The Japan Maritime Self-Defence Force (JMSDF) destroyers *Suzutsuki* (DD-117) and *Fuyuzuki* (DD-118) – respectively the third and fourth members of the *Akizuki* (DD-115) class of general-purpose destroyers – were both commissioned in March 2014. Together with *Akizuki* herself – commissioned in March 2012 – and *Teruzuki* (DD-116) – commissioned in March 2013 – they completed the four-ship class. The ships were conceived as a new type of 5,000-ton (standard displacement) general-purpose destroyer. They were the first of this type to be ordered by the JMSDF since as far back as the *Takanami* class (DD-110 to DD-114), which were commissioned between 2003 and 2006. In addition to undertaking missions envisaged for previous general-purpose destroyers, the new ships were conceived to meet the requirements of a new operational concept, explained in further detail below.

DESIGN BACKGROUND

The development of modern general-purpose destroyer designs for the JMSDF can be traced back to the Japanese Showa 52 year (1977), when the first member of the *Hatsuyuki* (DD-122) class was authorised.[1] Eventually extending to a class of twelve ships (DD-122 to DD-133) commissioned between 1982 and 1987, the ships marked a significant departure from previous Japanese escort designs. Notably, they added surface-to-surface and (short-range) surface-to-air missiles to the traditional gun-based armament found on older classes. Moreover,

Opposite: The lead Japanese *Akizuki* (DD-115) class destroyer pictured in April 2012, shortly after delivery. The latest in a long line of JMSDF general-purpose destroyers, the class incorporates improved local area air defence capabilities to protect specialist air-defence destroyers acting in the ballistic missile defence role. *(Japan Maritime Self Defence Force)*

The Japanese *Hatsuyuki* class general-purpose destroyer *Isoyuki* (DD-127) pictured at Portsmouth UK in July 2013 during a global training deployment. The *Hatsuyuki* class marked a significant departure from older Japanese destroyer designs, not least its use of surface-to-air and surface-to-surface missiles, as well as an integrated command system. It can be regarded as the start of an evolutionary design process from which the current *Akizuki* class can be traced. *(Alexander Waters)*

The *Akizuki* class is a further evolution of JMSDF general-purpose destroyer designs that include the *Hatsuyuki*, *Asagiri*, *Murasame* and *Takanami* classes. This series of images of *Akizuki* taken in April 2012 depicts the ship's overall configuration to good effect. A high freeboard hull is surmounted by two main superstructure blocks, with the two funnels in between corresponding to the arrangement of the two independent main engine compartments. The four paired panels of the FCS-3A multi-function radar are split between the forward and aft superstructure and arranged so as to provide 360° coverage. Much of the armament is concentrated forward of the bridge, although surface-to-surface missiles are located between the funnels. Although not an inherently stealthy design, the hull and sides are angled to limit radar reflections, the lattice mast of previous designs has been replaced by a stealthier structure and equipment such as torpedo tubes hidden within the ship's hull. *(Japan Maritime Self Defence Force)*

they were able to operate an anti-submarine (sea control) helicopter. Other enhancements included the installation of an integrated combat management system and the use of gas turbine propulsion. The class was the first in a series of general-purpose destroyer designs, viz.:

- *Asagiri* class: Eight ships (DD-151 to DD-158) commissioned between 1988 and 1991.
- *Murasame* class: Nine ships (DD-101 to DD-109) commissioned between 1996 and 2002.
- *Takanami* class: Five ships (DD-110 to DD-114) commissioned between 2003 and 2006.

Of these ships, the last two classes were noteworthy in using vertical launch systems (VLS) for both surface-to-air and anti-submarine missiles.

These four classes' integrated combat capabilities meant that they were well-equipped to expand the previous anti-submarine warfare orientation of older JMSDF escort vessels, being able to perform a full range of surface combatant missions to counter threats on, below or above the water. They can undertake operations in the direct defence of Japan, the surveillance of its surrounding waters and protection of vital trade routes. More recently, missions have expanded to encompass a broader range of international duties. These have included the support of Japanese contributions to global stabilisation and peacekeeping operations, as well as participation in anti-terrorist and anti-piracy task forces.

A significant development during this period has been the proliferation of various missile technologies, often associated with the changed environment following the end of the Cold War. Of these, the most concerning is, possibly, the much wider availability of ballistic missiles. It is a well-known fact that Japan itself is exposed to the risk of ballistic missile attack. To counter this threat, it was decided to apply upgraded air-defence facilities initially developed by the US Navy to the four Aegis-equipped *Kongou* (DDG-173) class air-defence destroyers commissioned between 1993 and 1998. More specifically, this included improvements to the detection and tracking capabilities of the ships' SPY-1D multi-function radar and changes to their weapons-control software to permit the launch of Standard SM-3 interceptor missiles to permit the destruction of ballistic missiles in the exo-atmosphere. The necessary work was undertaken under the FY2004–2007 defence budgets.[2] It was subsequently extended to the two new *Atago* (DDG-177) class Aegis-equipped destroyers that were commissioned in 2007 and 2008.

Although work on developing maritime-based countermeasures to the threat of a ballistic missile

The *Mursame* class destroyer *Ikazuchi* (DD-107) pictured off the Japanese coast in February 2016 – a snow-capped Mount Fuji is in the background. The overall configuration of the *Akizuki* class is similar to the preceding *Murasame* and *Takanami* class general-purpose destroyers but this resemblance hides some significant differences. In particular, the *Akizuki* class was developed to undertake a local area defence role in addition to retaining GP capabilities. *(US Navy)*

attack have therefore progressed steadily, the programme has resulted in the emergence of a significant vulnerability. Whilst the SPY-1D radar that is fitted to current Aegis-equipped ships has excellent capabilities to detect and track both high-altitude and low-altitude targets, there have been concerns over the extent of its ability to undertake both functions simultaneously. As such, an Aegis destroyer undertaking a ballistic missile defence (BMD) role might find itself less able to counter a direct, low-altitude attack, for example from a sea-skimming anti-ship missile.

It seems that the *Akizuki* class were developed with precisely this scenario in mind. In essence, the new destroyers are equipped to undertake an additional local area-defence role in addition to retaining general-purpose capabilities. As such, they are able to protect ships sailing in consort from air and missile attack as well as providing for their own point defence needs. This local force defence mission contrasts with the role of the specialised air-defence destroyers, which have a wider area-defence focus. This additional role has inevitably resulted in detailed changes in configuration and – particularly – equipment from previous general-purpose destroyer classes.

DESIGN CONFIGURATION

The new *Akizuki* class design is typified by the combination of an evolutionary approach to structural design with the incorporation of the latest weapons technology, particularly with respect to sensors and other electronics, as described below. Please refer also to Table 3.1.1 for additional detail.

Overall Configuration: The *Akizuki* class's overall hull design can be traced back to that adopted for the first time in the *Hatsuyuki* class more than thirty years ago and which has been followed – with variations – by all subsequent general-purpose destroyer classes. It provides for a high freeboard by means of a shelter deck arrangement that – in the *Akizuki* class – extends from bow to stern.[3]

In terms of overall layout, the main superstructure comprises a bridge forward, two funnel blocks and a hangar structure aft. There is also a helicopter landing area at the stern. Both the hull and superstructure are inclined to reduce the ship's radar signature. The main mast on top of the bridge has also adopted a stealth configuration – first seen in the *Atago* class – compared with the more traditional lattice structure used in most previous Japanese destroyer classes.

A notable characteristic is the arrangement of the arrays that form the class's principal radar system. The FCS-3A multi-function radar has replaced the conventional rotating arrays found on previous general-purpose destroyer designs. This new system – also found in an earlier version on the *Hyuga* (DDH-181) class helicopter-carrying destroyers – incorporates four sets of paired fixed panels of two different sizes.[4] Two sets are mounted at 04 deck level on the forward superstructure to cover forward port and starboard quarters; two further sets are located at the same level on the aft superstructure to cover the corresponding areas aft.

Much of the principal armament is located forward of the bridge. This includes a single-barrelled 127mm Mk 45 gun, a 32-cell Mk 41 VLS and a Phalanx close-in weapons system (CIWS), the latter mounted on a raised platform ahead of the bridge superstructure. Surface-to-surface missile launchers are located in the area between the two funnels and there is an additional CIWS mount on top of the hangar. There are twin torpedo tubes port and starboard in the forward superstructure.

The class's hangar can accommodate up to two helicopters. The flight deck incorporates a Mk 6 recovery, assist secure and traverse system (RAST). This is used to assist the recovery of a helicopter and

Two views of the second *Akizuki* class destroyer, *Teruzuki* (DD-116), pictured at speed. The class's propulsion system comprises four licence-built Rolls-Royce Spey SM1C gas turbines installed in a COGAG (combined gas and gas arrangement). These produce a maximum speed of 30 knots. *(Japan Maritime Self Defence Force)*

safely transfer it to the hangar. Ship's boats comprise an 11m work boat located to starboard of the forward funnel and a 7.6m fast rigid hulled inflatable boat (RHIB) located to port of the aft funnel. The boats are hoisted and recovered by means of hydraulic davits.

In the interior of the ship, the superstructure at No. 1 deck level houses officer accommodation. The combat information centre (CIC), communications room, machinery control room, galley and recreational spaces – as well as additional accommodation – are located on No. 2 level, connected by a through-deck passage. The machinery and equipment spaces, as well as magazine and storage areas are located at lower levels, along with some additional accommodation space.

Propulsion: The propulsion system installed in the *Akizuki* class destroyers comprises four sets of Rolls-Royce Spey SM1C gas turbines, which are used under a combined gas and gas (COGAG) arrangement. The turbines drive two shafts fitted with variable-pitch propellers. The propellers are five-bladed skew types, chosen so as to reduce underwater noise. The designed maximum speed is 30 knots.

There are two main engine compartments, each containing two gas turbines and associated reduction gear. Turbines No. 1 and No. 2, which drive the port shaft, are installed in the forward (No. 1) engine room, whilst turbines No. 3 and No. 4, driving the starboard shaft, are located in the aft (No. 2) engine room. One consequence of this arrangement is that the port shaft is longer than the aft shaft. There is a funnel above each of the two engine rooms, from which emissions from the gas turbines are emitted by means of ducts. The two engine rooms are divided from each other by an auxiliary machine room, thereby reducing the risk of loss of propulsion from a single hit. In similar fashion, the three electrical generators that provide power for ship 'hotel' functions are split between the two main engine compartments and the electrical generator room, thereby further enhancing survivability.

As previously mentioned, the machinery control room is located at No. 2 deck level. It contains the main consoles from which ship propulsion, generation and electricity distribution are remotely controlled by means of a platform management system. The machinery control room also acts as the principal damage control centre, acting as a communications hub from which damage such as fire, flooding, gas escape and bulkhead breakdown can be dealt with promptly and effectively.

ELECTRONIC OUTFIT

The equipment installed on the *Akizuki* class destroyers has been designed to allow them to combine the previous general-purpose destroyer role with the new 'local area defence' mission envisaged for the new design. To meet this combined requirement, the ships are equipped with some of the most sophisticated and modern technology developed by the Japanese defence sector.

Combat Management System: The heart of the *Akizuki* class's war-fighting capabilities is provided by its combat management system (CMS). This integrates the ship's various sensors, communications systems and weapons to allow it to conduct its role as a major surface combatant to maximum effect. The deployment of indigenous combat management systems in JMSDF general-purpose destroyers can be traced to the OYQ-5 tactical information processing and combat direction system, which was first installed in the *Hatsuyuki* class.[5] The OYQ series has since been progressively upgraded to take account of progress in both digital and communications technology, including the use of commercial off the shelf (COTS) equipment.

The OYQ-11 system found in the *Akizuki* class is a direct development of the OYQ-10 system installed in the *Hyuga* class. It has a fully distributed architecture and is linked to the ship's various sensors and weapons by the NOYQ-1B integrated

Table 3.1.1.

AKIZUKI (DD-115) PRINCIPAL PARTICULARS

Building Information:	
Fabrication Commenced:	17 July 2009
Launched:	13 October 2010
Commissioned:	14 March 2012
Builders:	Mitsubishi Heavy Industries (MHI) at its Nagasaki shipyard, Japan.
Dimensions:	
Displacement:	5,050 tons standard displacement, 6,800 tons full load displacement.
Overall Hull Dimensions:	151.0m x 18.3m (maximum) x 5.4m. Depth is 10.9m.
Equipment:	
Missiles:	4 x Mk 41 8-cell VLS modules for a total of 16 quad-packed ESSM surface-to-air missiles and 16 ASROC anti-submarine rockets. 2 x quadruple launchers for SSM-1B surface-to-surface missiles.
Guns:	1 x 127mm Mk 45 Mod 4 62-calibre gun. 2 x 20mm Phalanx CIWS. 12.7mm machine guns.
Torpedoes:	2 x triple HOS-303 324mm anti-submarine torpedo tubes for Type 97 lightweight torpedoes.
Aircraft:	Space for two medium (10-ton class) or one heavy (15-ton class) helicopters. 1 x SH-60K sea control helicopter normally embarked.
Countermeasures:	NOLQ-3D electronic warfare suite, including both ESM and ECM. 4 x Mk 137 launchers for SRBOC Mk 36. Type 4 towed decoy system. OLQ-1 torpedo defence system, including Type 1 Mobile Decoy (MOD) and Floating Acoustic Jammer (FAJ).
Principal Sensors:	1 x FCS-3A multifunction radar. 1 x OPS-20C navigation radar. 1 x OQQ-22 integrated sonar suite, including OQR-3 towed array.
Combat System:	OYQ-11 advanced combat direction system. Comprehensive communications system includes Links 11 and 16.
Propulsion Systems:	
Machinery:	COGAG. 4 x RR Spey SM1C gas turbines each rated at c. 12MW total produce up to 64,000shp through two shafts.
Speed:	Designed maximum speed is 30 knots.
Other Details:	
Complement:	A typical crew comprises c. 210 personnel.
Class:	Four ships have been constructed: *Akizuki* (DD-115), *Teruzuki* (DD-116), *Suzutsuki* (DD-117) and *Fuyuzuki* (DD-118).

Note:
The ships completed after *Akizuki* have a slightly higher standard displacement and are equipped with the indigenous Type 07 variant of ASROC.

ship-wide network. The combination of OYQ-11 and various components such as the FCS-3A fire-control system, the OQQ-22 sonar system and the NOLQ-3D electronic warfare system provides functionality similar to the US Navy's Aegis. Operator interface is by means of multi-function UYQ-70 consoles. The system incorporates a high level of automation, based on a series of pre-programmed war-fighting rules, to speed response times.[6]

Radar Equipment: The principal radar system found onboard *Akizuki* and her sisters is the multi-function FCS-3A, sometimes claimed to be the most advanced radar system in the world. An example of the latest generation of active phased arrays, the FCS-3A is the core component of the Type 3A anti-air warfare defence system. As previously mentioned, the radar itself is an improved version of the FCS-3 system fitted to the *Hyuga* class.

The main radar system on the *Akizuki* class is the indigenous FCS-3A, a development of the FCS-3 system installed in the *Hyuga* class. It comprises four pairs of fixed panels, one pair above the bridge being shown here. *(Tomohiko Tada)*

Initial research and development work on the

***Akizuki* (DD-115)** (2012)
1:700 scale

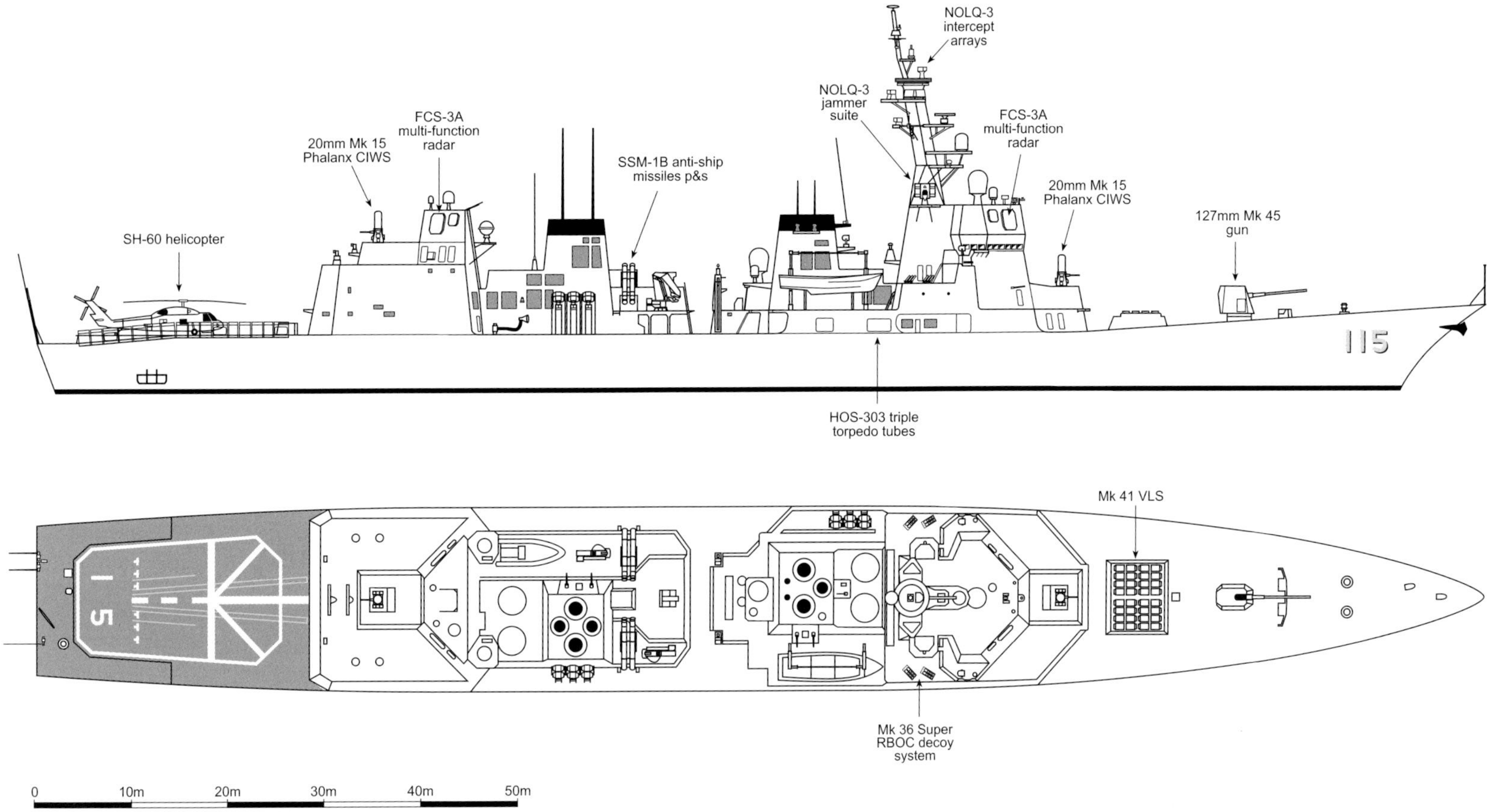

(Drawings © John Jordan, 2016)

Fuyuzuki (DD-117) entering harbour in wintry conditions; the helicopter-carrying destroyer *Hyuga* (DDH-181) can be seen undergoing maintenance in the background. The image shows the ship's stern arrangement, including the trumpet-like horns for the Type 4 towed torpedo-defence array to starboard. Also evident is the asymmetrical arrangement of the hangar doors and the rear panels for the FCS-3A radar system. *(Japan Maritime Self Defence Force)*

FCS-3 system was initiated by the Technical Research and Development Institute (TRDI) of the Japan Defence Agency (now the Japanese Ministry of Defence) during the 1980s. A development prototype was subsequently mounted on the trials ship *Asuka* (ASE-6102) during the 1990s. Production of the radar was assigned to Mitsubishi Electric Corporation.

As installed in the *Hyuga* class, the FCS-3 system includes four large antennae mounted on the superstructure to provide 360° coverage in similar fashion to the SPY-1 panels associated with the US Navy's Aegis. Each of the four arrays operates in the 4,000–8,000 MHz US Navy C (NATO G/H) band and comprises numerous transmitting/receiving (Tx/Rx) modules formed from gallium arsenide

A detailed view of the forward superstructure of *Teruzuki* showing the paired arrays of the FCS-3A radar system. The larger arrays work in the NATO G/H bands to search for and track potential targets; the smaller arrays undertake fire control functions and operate in the higher definition NATO I/J bands. *(Tomohiko Tada)*

This view of the forward superstructure of *Teruzuki* shows some of the countermeasures equipment fitted to the class. The ESM elements of the NOLQ-3D electronic warfare suite are fitted on platforms towards the top of the mast whilst two hexagonal box-like ECM jammers are fitted on sponsons at the mast's base. Also visible at a lower level are two six-barrelled Mk 137 launchers for the Mk 36 SRBOC decoy system. *(Tomohiko Tada)*

(GaAs) semiconductors. These arrays are used to form electronically-controlled beams to search for, detect and track potential targets.[7] There are also four smaller arrays, operating in the 6,000–11,000 MHz US Navy X band (NATO I/J band), that are used as fire-control radars for the semi-active homing Evolved Sea Sparrow missiles (ESSM) that comprise the principal air-defence armament.[8]

The FCS-3A variant of the system installed in the *Akizuki* class incorporates significant enhancements over the earlier FCS-3 version, extending its detection and tracking range whist improving its performance in overcoming sea- and land-based clutter when operating in coastal waters. The improvements to range were achieved by exchanging the conventional GaAs modules used in the original system for semiconductors formed from gallium nitride (GaN) to increase output power. The enhancements in coastal performance are due to the development of enhanced signal processing algorithms. Additional software improvements also allow the FCS-3A system to control the ship's main 127mm gun as well as its surface-to-air missiles.

Given the primary radar system's multi-function capabilities, there is no need for additional arrays such as those for surface search. However, an OPS-20C radar system is shipped for navigational purposes and this can also be used to monitor surface targets at close range. It comprises two small rotating antennae – a main array and a smaller subsidiary antenna – located on the mast.

Electronic Warfare Equipment: The *Akizuki* class's integrated NOLQ-3D electronic warfare suite encompasses both electronic support measures (ESM) and electronic countermeasures (ECM). The main ESM detection equipment is positioned towards the top of the mast, with two ECM jammers located to port and starboard of its base at No. 5 deck level. The system is complemented by two pairs of six-barrelled Mk 137 decoy launchers for the Mk 36 Super Rapid Bloom Offboard Countermeasures (SRBOC) decoy launching system, which can fire both radio wave distorting chaff and infrared decoys.

Sonar Equipment: In addition to their air-defence role, the *Akizuki* class's general-purpose configuration also means that the ships must be well-equipped for anti-submarine warfare. This requires incorporation of both a hull-mounted and towed array sonar in the design. The hull sonar installed in the class is an OQQ-22, a modified version of the OQQ-21 installed in the *Hyuga* class. A large sonar with both active and passive functionality and capable of operation in a range of frequencies, the OQQ-21 was also tested on the *Asuka* before installation in operational vessels. The OQQ-22 variant installed in the *Akizuki* class lacks its predecessor's flank arrays but is paired with an OQR-3 towed array, a development of the system found in previous general-purpose destroyers. The sonar system's processors can also handle signals transmitted, for example, by sonobuoys deployed by the ship's anti-submarine helicopters.

The *Akizuki* class also benefits from a torpedo-defence system that is integrated with the sonar system. The new OLQ-1 torpedo-defence system comprises both the Type 1 self-propelled Mobile Decoy (MOD) and the Floating Acoustic Jammer (FAJ). The former is deployed by a quadruple MOD launcher similar to a lightweight torpedo-tube mounting that is enclosed within the ship's superstructure. Launch is by means of compressed air. Each decoy has a diameter of 15.2cm, a length of 100cm and weighs approximately 25kg. The decoy emits an acoustic signal to lure approaching torpedoes away from the ship. The FAJ is deployed from a single launcher on the centreline forward of the second funnel. The launcher can train 130° to port and starboard, has between five and eighty degrees elevation and a maximum projection range in the order of 1km. The decoy falls into the sea on a parachute after flying a pre-determined range and, again, attempts to distract incoming torpedoes by means of an acoustic signal. The *Akizuki* class is also equipped with the Type 4 torpedo-defence towed array that is already operated by the *Hyuga* class and some other ships. It is deployed from two trumpet-like openings on the starboard side of the stern.

Communications Equipment: As might be expected, a comprehensive range of communications equipment is installed in the *Akizuki* class. Satellite communications systems include antennae for the X band NORA-1C and Ku Band NORQ-1 networks. These connect, respectively, with the Superbird B2 and Superbird C2 satellites. In each case, the relevant equipment is housed in spherical radomes located on the mast and superstructure. In addition, USC-42 satellite communications antennae are fitted to facilitate joint operations with the United States Navy. A further system – the NORC-4B – provides services such as voice telephone, facsimile and data connectivity through the Inmarsat commercial network. The ships are also equipped with a specialised helicopter data link system – ORQ-1C-2 – that is used to exchange data, for example with respect to anti-submarine operations, between the ship and helicopters in flight. The relevant equipment is housed in a radome fitted to a platform towards the top of the mast.

In addition to satellite communications systems, several whip aerials are mounted for radio communications. These are most frequently used to support tactical digital information links, most notably the Link 11 and Link 16 systems used by the United States and its allies.

WEAPONRY

As for the class's electronics systems, the *Akizuki* class's weaponry is amongst the most modern deployed by the JMSDF. Much of it reflects the strong US Navy influence that is present amongst much JMSDF equipment, albeit a steady process of indigenisation and adaptation to local requirements is also evident.

Guns: The main gun – mounted on the class's foredeck – is a 127mm Mk 45 Mod 4 62-calibre single-barrelled weapon, the latest iteration of a gunnery system first developed for the US Navy in the 1960s. The barrel is eight calibres – roughly 1m – longer than the US Navy's older 54-calibre 127mm weapons and the Italian company Oto Melara's equivalent weapon. The additional length makes the gun more suitable for shore bombardment than previous weapons in the series, extending range from 13nm to 20nm (c. 24km to c. 36km). It also allows for greater precision.

A detailed view of the aft superstructure of *Teruzuki*, showing the rear FCS-3A radar panels, the aft Phalanx close-in weapons system and some of the equipment used to facilitate helicopter landings. The class is fitted with two Mk 15 Phalanx systems, which are located forward and aft so as to provide 360° coverage. *(Tomohiko Tadu)*

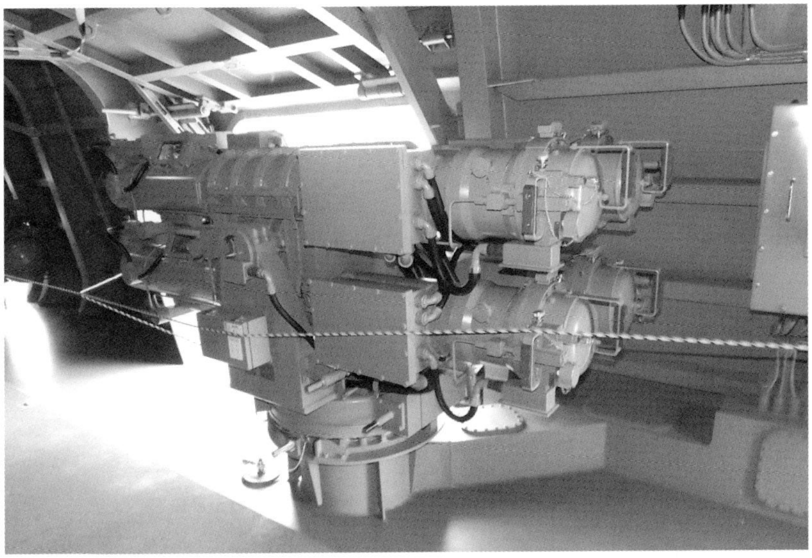

In addition to a Type 4 towed torpedo-defence array, the *Akizuki* class are equipped with a new torpedo-defence system incorporating both the Type 1 self-propelled MOD mobile decoy and the FAJ floating acoustic jammer. These images show the torpedo tube-like launcher for the MOD and the mortar-like launcher for the FAJ. *(Tomohiko Tada, Yasu Osugi)*

Akizuki and her sisters are also fitted with two Mk 15 Phalanx close-in weapons systems, which combine a rotating, six-barrelled 20mm Gatling gun with a small Ku band (12,000–18,000 MHz) target acquisition and tracking radar. One mount is located forward of the bridge and one on the hangar roof to provide 360° coverage. Phalanx was originally developed to provide point-defence against anti-ship missiles but the latest Block 1B variant also has enhanced capabilities against small surface vessels, as well as other asymmetric threats, such as drones.

The class is also capable of using 12.7mm machine guns from mountings that have been prepared on the sponsons that extend from either side of the bridge.

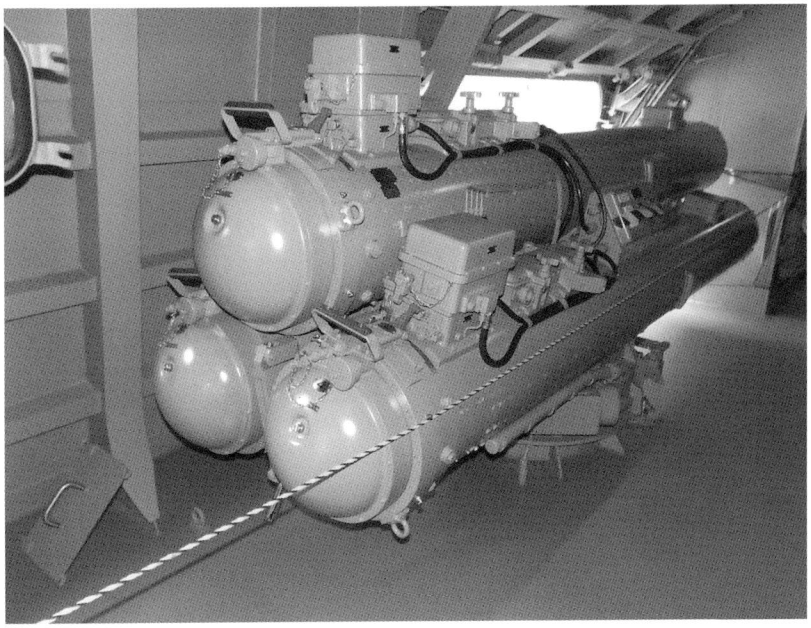

Although they have an important local area air-defence role, the *Akizuki* class retain the powerful anti-submarine capabilities of previous Japanese general-purpose destroyer designs. In addition to embarked helicopters and torpedoes deployed by ASROC rockets, anti-submarine torpedo tubes are also shipped for short-range defence. These images provided a detailed view of the triple HOS-303 torpedo-tube mount and an external view showing the mount's enclosed position within the forward superstructure. The opening on the far left is for the quadruple MOD launcher that forms part of the torpedo defence system. *(Tomohiko Tada)*

Much of the *Akizuki* class destroyers' principal armament is located forward of the bridge, including a long 127mm Mk 45 Mod 4 62-calibre gun and a 32-cell Mk 41 VLS for ESSM and ASROC missiles. These systems can be seen clearly in this picture of *Teruzuki* entering San Diego on 15 October 2014. *(US Navy)*

Guided Weapons: The *Akizuki* class's guided weapons systems include anti-aircraft, anti-submarine and anti-surface missiles. Of these weapons, the class's surface-to-air missile – crucial for the local force defence mission – is the RIM-162 Evolved Sea Sparrow Missile (ESSM). This was developed by the Raytheon Company for the United States and allied navies. A complete redesign of the original RIM-7 Sea Sparrow series used by the JMSDF since the 1980s, ESSM has a range in excess of 30km, making it suitable for local area defence requirements. On *Akizuki* and her sister-ships, ESSM is fired from the Mk 41 VLS located forward of the bridge. Sixteen of the launcher's thirty-two cells are nominally allocated to ESSMs. Each of these cells is capable of housing four quad-packed missiles, giving a total load-out of sixty-four missiles. In addition to its improved range, ESSM benefits from significant enhancements to guidance technology over its predecessor, including various target illuminations options dependent on the nature of the target and the tactical situation.

The VLS also houses vertically-launched ASROC (Anti-Submarine Rocket) missiles. The nominal outfit is sixteen missiles on the basis of one missile per cell. ASROC is a lightweight torpedo deployed by a solid-propellant rocket that is fired towards the sea area where a submarine detected by sonar is thought to be located. The torpedo separates from the rocket when the target area is reached, descending to the sea by means of a parachute. The torpedo operates as a homing weapon once it has been launched into the sea. The JMSDF originally used the US Navy's ASROC system but – from *Teruzuki* (DD-116) onwards – has now standardised on an indigenous Type 07 torpedo-projection rocket. Similarly, the weapon deployed by the ASROC system is the indigenous Type 97 torpedo.

Indigenisation has also progressed to the anti-ship missile armament carried by the class. Eight Type 90 SSM-1B are mounted in two quadruple launchers installed immediately forward of the second funnel. The SSM-1B missile has a turbojet engine and its maximum range has been reported as being in excess of 100km. The missile is closely related to similar weapons deployed from Japan Self-Defence Forces' land vehicles and aircraft.

Underwater Weapons: The principal anti-submarine weapon deployed on the class is the Type 97 lightweight torpedo. In addition to being utilised in vertically-launched ASROC missiles, these can also be launched from triple HOS-303 mountings that are located in both sides of the superstructure just forward of the first funnel. It can also be deployed by an embarked sea control helicopter.

The Type 97 weapon will soon be replaced by the new Type 12 torpedo, which was developed from the late 1990s onwards by the TRDI under the designation GRX-5.[9] The new torpedo has been specifically designed to have an improved performance in littoral waters, where there is greater variability in areas such as temperature, water flow, salinity and seabed shape. It retains the propulsion system and warhead of its predecessor but incorporates a new sensor to deal more effectively with these conditions. The signal processing capabilities of the guidance system have also been improved.

Helicopters: The *Akizuki* class ships incorporate a helicopter hangar that is capable of accommodating two medium helicopters (c. 10-ton class) or one larger helicopter (c. 15-ton class). These correspond to the JMSDF's SH-60K sea-control and MCH-101 mine countermeasures and transport helicopters. Either of these types could be embarked depending on operational requirements but the usual complement is one SH-60K helicopter.

Based on the US Navy's Sea Hawk series, the SH-60K was developed as a successor to the SH-60J, with procurement commencing in FY2002. Sensors include a low-frequency dipping sonar, a forward-looking infrared search and tracking system and an inverse synthetic aperture radar. The main anti-submarine weapons shipped are indigenous Type 97 lightweight torpedoes. These are supplemented with Hellfire air-to-surface missiles and 7.62mm light machine guns to counter a spectrum of hostile surface targets. The SH-60K is also equipped with the Ship Landing Assist System (SLAS). This allows automated landing on the flight deck so as to reduce the burden on a helicopter's pilot during landings at night or in bad weather.

The MCH-101 is based on the Finmeccanica Helicopters (formerly AgustaWestland) AW-101 series. Procurement of the type started in 2003 as a successor to the MH-53E Sea Dragon. In mine-hunting configuration, it is equipped with a towed sonar and laser scanner to facilitate high-speed searches of suspicious objects.

CONSTRUCTION AND DEPLOYMENT

The four ships of the *Akizuki* class were authorised under the FY 2007, FY2008 and FY2009 construction programmes, with two ships being approved in the final year. The lead ship – DD-115 – cost JPY 74,972m or approaching US$700m at then current exchange rates (US$1 = JPY 110). The budget for the next ship amounted to JPY 68,987m, whilst a total amount of JPY 145,101m was authorised for the final pair.[10] Construction of the first three ships was allocated to Mitsubishi Heavy Industries' Nagasaki shipyard, whilst DD-118 was built by Mitsui Engineering & Shipbuilding of Tamano. Assembly of all the ships was completed within three years; please refer to Table 3.1.2 for key construction dates.

On commissioning, one member of the class has

A view of *Akizuki* taken around the time of her commissioning. She was authorised under the FY2007 construction programme at a cost of c. US$700m and delivered in 2012. *(Japan Maritime Self Defence Force)*

Table 3.1.2: *AKIZUKI* CLASS LIST

NAME	PENNANT	ORDERED	COMMENCED	LAUNCHED	COMMISSIONED	BUILDER
Akizuki	DD-115	FY2007 Programme	17 July 2009	13 October 2010	14 March 2012	Mitsubishi Heavy Industries Ltd, Nagasaki
Teruzuki	DD-116	FY 2008 Programme	2 June 2010	15 September 2011	7 March 2013	Mitsubishi Heavy Industries Ltd, Nagasaki
Suzutsuki	DD-117	FY 2009 Programme	18 May 2011	17 October 2012	12 March 2014	Mitsubishi Heavy Industries Ltd, Nagasaki
Fuyuzuki	DD-118	FY 2009 Programme	14 June 2011	22 August 2012	13 March 2014	Mitsui Engineering & Shipbuilding Co Ltd, Tamano

An image of *Akizuki* after her launch on 13 October 2010. She was built at Mitsubishi Heavy Industries' Nagasaki Shipyard, which was responsible for constructing three out of four members of the class. *(Japan Maritime Self Defence Force)*

been allocated to each of the JMSDF's four, eight ship-strong escort flotillas. These flotillas are subdivided into two equally-sized escort squadrons, with the *Akizuki* class serving in the same squadrons as the *Kongou* class destroyers that are allocated to the ballistic missile defence role. In practice, the ships often deploy independently or as part of smaller groups; they have already operated as far away as the Indian Ocean and off the United States' west coast.

Priority given to the expensive task of replacing the first-generation helicopter-carrying destroyers with the new *Izumo* (DDH-183) class has meant that it was not until FY2013 that further general-purpose destroyers were authorised for the JMSDF. The design of the new 25DD Type – which also carries the 5,000-ton type destroyer designation used for the *Akizuki* class – continues the evolutionary approach adopted in previous general-purpose destroyer classes. Similarities include continued use of the basic *Akizuki* hull design and ongoing specification of the FCS-3 series as the main radar system. However, it appears that the new class will have a greater orientation towards anti-submarine warfare than the *Akizuki* class. This is reflected in the specification of an innovative combined gas turbine-electric and gas turbine (COGLAG) propulsion system, as well as further enhancements to the sonar suite. It is also hoped that use of COGLAG propulsion will reduce overall life-cycle costs, with the emphasis on economy also reflected in a reduction in the Mk 41 VLS to only sixteen cells and concentration of the FCS-3 arrays on the forward superstructure. The authorised cost

Suzutsuki and *Teruzuki* pictured fitting out at Mitsubishi Heavy Industries' Nagasaki Shipyard in December 2012. Work on *Teruzuki* was relatively well advanced at this stage; she was commissioned on 7 March of the following year. *(Luck One)*

Fuyuzuki (DD-118) – the last member of the *Akizuki* class to be completed – seen operating with the US Navy aircraft carrier *Theodore Roosevelt* (CVN-71) and the Indian replenishment tanker *Shakti* during the Malabar 2015 exercises in the Indian Ocean. The *Akizuki* class has already been deployed quite widely as the JMSDF continues to expand international engagements.

of the first unit was some JPY 70.1bn, somewhat less than *Akizuki*. Construction has been entrusted to Mitsubishi Heavy Industries at Nagasaki, with completion scheduled for March 2018. A second ship was included in the FY2014 defence budget.

CONCLUSION

The *Akizuki* class general-purpose destroyers are now in operational service and ready to perform the extended local area defence mission that is inherent in the new design. The largest and most-capable DD-type destroyers yet to serve in the JMSDF, they are an excellent example of the evolutionary design approach that has achieved significant cumulative progress since the commissioning of the first of the *Hatsuyuki* class a little over thirty years ago.

Notes:

1. The *Akizuki* class destroyers were authorised in the Heisei 19 through to Heisei 21 years so they are referred to as 19DD, 20DD, 21DD and 21DD by the JMSDF. The Japanese Heisei era corresponds to the reign of Akihito, the current Emperor of Japan, which commenced on 8 January 1989. Accordingly, Heisei 19 corresponds to 2007. The previous Showa era relates to the reign of the former Emperor Hirohito between 25 December 1926 and 7 January 1989.

2. Launch tests of SM-3 Block IA interceptor missiles following completion of the necessary upgrade works were carried out in consecutive years from 2007 in the order of *Kongou* (DDG-173), *Chokai* (DDG-176), *Myoko* (DDG-175) and *Kirishima* (DDG-174). The tests were known as Ship Qualification Trials (SQTs) and allocated the numbers JFTM-1 to JFTM-4. During these tests, *Kongou, Myoko* and *Kirishima* were successful in shooting down simulated missile targets. The trial involving *Chokai* was unsuccessful because of problems with the guidance system of the LEAP (Lightweight Exo-Atmospheric Projectile) homing vehicle used in the final interception stage. For further details of the broader US Navy-led ballistic missile defence programme please refer to Norman Friedman's 'Ballistic Missile Defence and the USN' in *Seaforth World Naval Review 2013* (Barnsley: Seaforth Publishing, 2012), pp.184–91.

3. The *Akizuki* class carry a particularly distinctive resemblance to the preceding *Murasame* and *Takanami* classes. However, in addition to being around 1.5m broader than their predecessors, the new ships pay far more attention to stealth than older general-purpose destroyers, including a greater inclination of their hull and superstructure and expanded use of covered deck areas. Another significant difference is the use of propulsion machinery standardised on Rolls-Royce Spey gas turbines compared with the mixed Rolls-Royce/GE gas turbine plant of the previous two general-purpose destroyer classes.

4. The *Hyuga* class was described in the author's '*Hyuga* Class DDH-Type Destroyers: Japan's New Through-Deck Surface Combatants', *Seaforth World Naval Review 2014* (Barnsley: Seaforth Publishing, 2013), pp.88–103. It is interesting to note the considerable commonality in equipment between the two classes in spite of their significantly different roles.

5. The best English language summary of the development of the various combat management systems in the OYQ series is provided in Norman Friedman's *The Naval Institute Guide to World Naval Weapon Systems – Fifth Edition* (Annapolis MD: Naval Institute Press, 2006), p.78. The first systems – designated OYQ-1 and OYQ-2 were installed in the air-defence destroyers *Tachikaze* (DDG-168) and *Asakaze* (DDG-169) in the 1970s. The series has gone through several iterations subsequently.

6. The combination of OYQ-10 combat management system and the various other warfighting systems on the *Hyuga* class is known as the Advanced Technology Combat System (ATECS). ATECS is also often used to describe the similar combination found onboard the *Akizuki* class.

7. The system installed in the *Hyuga* class is also utilised for helicopter flight control.

8. FCS-3 was originally designed to operate with a surface-to-air version of Japan's AAM-4 active radar guided air-to-air missile, which carries its own active seeker for terminal guidance. When this variant was cancelled, the JMSDF had to develop a supplementary illumination system for use with the semi-active ESSM, which needs external radar illumination of the target in an engagement's final phase. It has been widely reported that Japan acquired the intermittent continuous wave illumination (ICWI) sub-system developed by Thales Nederland for their APAR multi-function radar for this purpose.

9. The Type 97 torpedo was previously developed under the designation GRX-4.

10. Data from the White Papers *Defense of Japan* for 1997 through to 1999 (Tokyo: Japanese Ministry of Defence).

3.2 SIGNIFICANT SHIPS

Author:
Guy Toremans

DOKDO CLASS ASSAULT SHIPS

SOUTH KOREA'S LPX-1 DESIGN

Since its initial foundation in 1948, the Republic of Korea Navy (ROKN) has seen a steady growth in size and capability. Initially equipped with surplus US Navy vessels, the navy saw the first indigenous-built units inducted into the fleet during the 1970s. The next decade saw the arrival of locally-constructed major surface combatants in the form of the *Ulsan* class frigates and the *Dong Hae* and *Po Hang* class corvettes, which that were largely equipped with 'off the shelf' US and European equipment. By the 1990s, progressively larger domestically-constructed surface combatants were joining the fleet, accompanied by efforts to grow an indigenous equipment industry. It was also during the 1990s that the ROKN's twentieth Commander-in-Chief, Admiral Ahn Byoung-Tae, achieved the necessary political support to develop the navy into a regional 'blue water' force, an ambition set out in his document *Navy Vision 2020.*[1] A key element of this vision was the strategic mobile concept: the creation of a 'high end' flotilla of ships capable of flexible, independent operations on the high seas. The need to maintain strong coastal forces to counter potential North Korean aggression has acted as something of a restraint on the extent to

Two images of the Republic of Korea Navy's amphibious assault ship *Dokdo* operating in company with the US Navy's *Wasp* class LHD *Bonhomme Richard* (LHD-6) and other amphibious ships on 27 March 2014. The origins of the *Dokdo* class can be traced to the achievement of political support during the 1990s to develop the Republic of Korea Navy's capabilities through the creation of strategic mobile forces capable of 'blue water' deployment throughout the Asian region. The centrepiece of these forces would be small aviation-capable ships. In practice, however, *Dokdo* was heavily influenced by amphibious requirements and is formally assigned to the navy's amphibious squadron. *(US Navy)*

which vision has been achieved. Nevertheless, the ROKN has successfully transitioned into a powerful regional navy, fully capable of international deployment and with highly sophisticated vessels that include Aegis destroyers and AIP-equipped submarines. Also prominent amongst these new warships are the *Dokdo* class amphibious assault ships, amongst the most potent of the type in the region.

GENESIS

One of the elements of the strategic mobile fleet concept was the construction of aviation-capable ships that could act as flagships for the fleet's deployed task groups. However – probably due to the ongoing threat from South Korea's northern neighbour and the need to maintain powerful littoral warfare capabilities to act as a deterrent – the ships that ultimately emerged were also heavily influenced by amphibious requirements. These requirements called for a platform capable of carrying an expeditionary force (including tanks, vehicles and support elements), as well as rotorcraft and landing craft for use in an amphibious assault. The design also needed to be configured to provide command and control capabilities across both the maritime and land domains, as well as having the ability to transport considerable volumes of

Table 3.2.1.

DOKDO (6111) PRINCIPAL PARTICULARS

Building Information:	
Fabrication Commenced:	February 2003
Launched:	12 July 2005
Commissioned:	3 July 2007
Builders:	Hanjin Heavy Industries & Construction Co. Ltd., Busan, Republic of Korea
Dimensions:	
Displacement:	c. 14,000 tons standard displacement, 18,860 tons full load displacement.
Overall Hull Dimensions:	199.5m x 31.5m x 6.5m.
Equipment:	
Aircraft:	Typical air group of up to 10 medium-sized helicopters. There are five operating spots on the flight deck, one for heavy helicopters.
Missiles:	1 x 21-cell Mk 49 RAM launch system for RIM-116B Block 1 surface-to-air missiles.
Guns:	2 x Thales Goalkeeper 30mm CIWS.
Countermeasures:	Nex-1 Future SLQ-200(V) Sonata electronic warfare suite. 2 x Dagaie Mk 2 decoy launchers.[1]
Principal Sensors:	Thales SMART-L long-range search radar; Thales MW-08 search and target-indication radar, GECO Engineering SPS-95K search and navigation radar. Selex Galileo SPN-720 approach radar, navigation radar.
Combat System:	Thales TACTICOS-derived combat management system. UNIMO Technology communications system includes Links 11 and 16.
Propulsion Systems:	
Machinery:	CODAD. 4 x Doosan Heavy Industries/MAN Diesel 16 PC 2.5 STC diesel engines each rated at c. 7.65MW deliver 41,615shp through two shafts. 1 x bow thruster.
Speed:	Designed maximum speed is 23 knots. Range is 10,000nm at 12 knots.
Other Details:	
Complement:	A typical crew comprises c. 245. Accommodation is provided for c. 420 plus embarked forces.
Military Lift:	Can accommodate c. 720 troops (c. 1,000 in overload conditions). Significant vehicle and cargo-carrying capacity. Space in well deck for 2 x LSF-II LCAC, with alternative LCM, LCU and AAV loadouts.
Class:	*Dokdo* (6111). A second unit – *Marado* – which will incorporate improvements is under construction.

Notes:

1 Many sources state *Dokdo* is equipped with the SLQ-260X TACM torpedo countermeasures system but its presence is not clear from published images of the ship.

containerised supplies and provide medical support during non-combat operations. The resultant LPX (Landing Platform Experimental) programme took tangible form in October 2002, when a contract was signed with Busan-based Hanjin Heavy Industries & Construction to construct the first of a planned total of three ships following receipt of formal cabinet approval.[2]

The new ship was the largest amphibious warship ordered from an Asian yard at that time. The design incorporated a large flight deck, a well deck and sufficient accommodation for a battalion of marines. There was also space for a mix of helicopters, tanks/armoured vehicles and wheeled vehicles, as well as an appropriate level of self-sufficiency in terms of ammunition, fuel, water, provisioning and logistic support. Given a lack of previous local experience in designing ships of this type, specialist assistance was sought from the US naval architecture and engineering group John J McMullen Associates (now part of Alion Science & Technology). This company provided design support in the areas of electromagnetic compatibility, topside structures, flight-deck and aviation facilities, and the stern well deck. This may account for *Dokdo*'s superficial resemblance to the US Navy's much larger *Wasp* (LHD-1) class amphibious assault ships.

***Dokdo* (6111)** (2007)
1:1000 scale

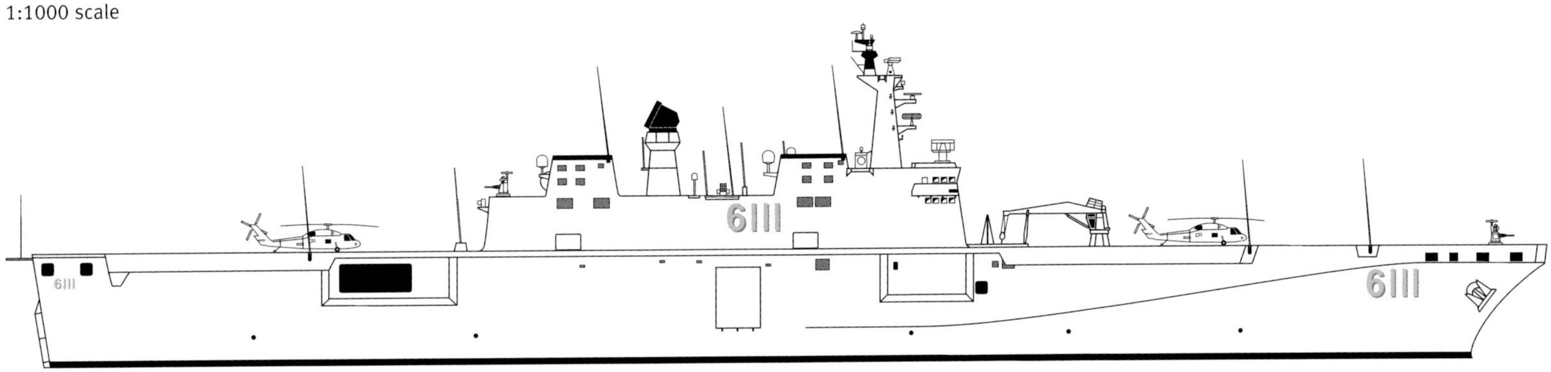

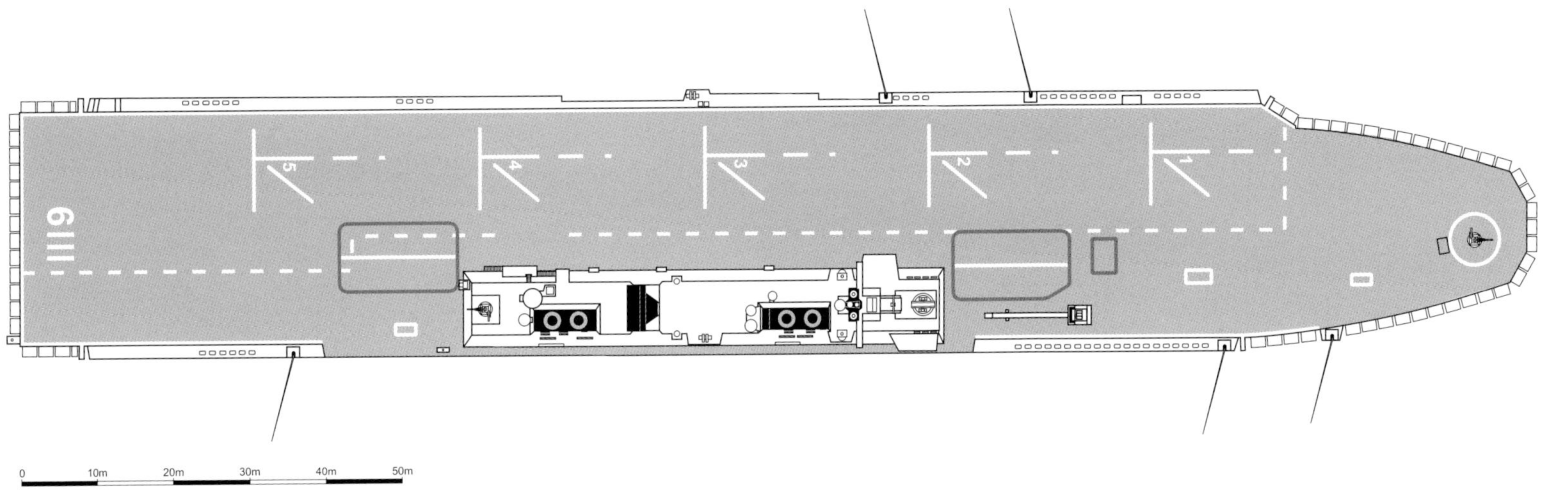

(Drawings © John Jordan, 2016)

Dokdo pictured at the Hanjin Heavy Industries & Construction shipyard in Pusan at the time of her official naming ceremony on 12 July 2005. The Thales SMART-L radar on her mainmast was a wooden mock-up. *(Republic of Korea Navy)*

The keel of the new ship was laid in February 2003. Named *Dokdo*, her launch was celebrated on 12 July 2005.[3] She was subsequently commissioned on 3 July 2007 at the naval base of Jinhae (also anglicised as Chinhae), taking the pennant number 6111. On her maiden deployment, the new ship took part in the biannual maritime defence exhibition LIMA 2007, held on Langkawi Island in Malaysia. Subsequently, on 1 January 2008, *Dokdo* was declared fully operational. It was originally hoped that a second ship would follow by c. 2010, with a third being completed later in the decade. However, this plan was significantly delayed. Consequently, it

Below: An image of *Dokdo* at the LIMA 2007 defence exhibition at Langkawi Island in Malaysia. *Dokdo* attended this exhibition soon after commissioning in July 2007. *(Guy Toremans)*

was only in December 2014 that the follow-on ship, named *Marado*, was ordered.

OVERALL DESIGN

Dokdo is a steel-built mono-hull ship with a length of c. 199.5m, a beam of c. 31.5m, a draught of c. 6.5m and a full load displacement of 18.860 tons. Full technical details are provided in Table 3.2.1.

Although the ROKN has been reluctant to release precise details on the ship's technical characteristics, payload and performance, it is understood that *Dokdo*'s military lift capability is built around a requirement to embark, transport, deploy and support an expeditionary force of around 720 troops and their equipment. This capacity can be increased to a force of around 1,000 personnel in 'overload' conditions. However, *Dokdo*'s capabilities go beyond this core specification, as she is effectively designed to deliver four capabilities from a single hull, acting as:

- An amphibious platform.
- An aviation support vessel.
- A joint command platform.
- A disaster-relief and hospital facility.

The need to support these varying capabilities has driven a polyvalent design, which has considerable ability to be 'flexed' between missions. This is reflected in an internal arrangement focused on a number of payload areas encompassing (i) the flight deck, (ii) a hangar and garage facilities, including a linked well deck and (iii) accommodation both for the ship's crew and the forces that may need to be embarked for any given mission.

The flight deck provides five landing spots and is capable of operating all helicopters currently in the South Korean Armed Forces' inventory. The after-most landing spot has greater clearances than the others and can be used by larger rotorcraft such as the CH-53 Sea Stallion series or the MV-22 Osprey. Some sources suggest the flight deck is strengthened and coated with special urethane to resist heat generated from aircraft, thus allowing operation by vertical/short take-off and landing (V/STOL) aircraft should the need arise. The ship's island is located amidships on the starboard side and houses a roomy navigation bridge, an admiral's staff bridge and a flight control centre with an excellent view of the flight deck. The majority of the ship's sensors and communications equipment are mounted on top of the island superstructure.

Accommodation facilities – many on gallery decks to either side of the main garage/hangar space – provide the crew with comfortable, noise-protected cabins, as well as recreation and living spaces, toilets, messes, a galley and offices. Working and accommodation areas are arranged in a logical structure to provide easy access and the passageways and stairways inside the ship are wide enough to allow fully-loaded combat troops to move to their assault launch points without difficulty. The ship's permanent complement numbers c. 245 crew but total accommodation is provided for c. 420. This would allow, for example, a command staff of around 150 personnel to be embarked. Senior officers have single cabins while other officers share either double or quadruple cabins; petty officers are accommodated in four- to six-berth cabins, whilst junior ratings share twelve-berth facilities. There is additional berthing for the 720 troops that comprise the normal expeditionary force. Having several hundred troops onboard puts a premium on maximising usable internal volume, cargo space and

Another image of *Dokdo* at the LIMA 2007 defence exhibition at Langkawi Island in Malaysia. Nearly 200m in length and displacing some 18,860 tons at full load displacement, her design is significantly flexible to facilitate the performance of a variety of roles. *(Guy Toremans)*

A distinguishing feature of *Dokdo* is a flexible garage/hangar which – including the well deck – runs for most of the ship's length. It can be accessed by elevator from the flight deck, from a side vehicle ramp or from the well deck itself. This view is taken forwards, looking aft towards the well deck. *(Guy Toremans)*

accommodation quarters and the embarked troops' quarters are inevitably more austere than those provided for the normal crew.

A distinguishing feature of the ship is the large hangar/garage and well deck area, which runs for most of the ship's length. The stern well deck is around 800m^2 and can be operated in either dry or wet status to accept either two landing craft air cushion (LCAC) or other LCM and LCU landing craft load-outs.[4] Ballasting the well dock is controlled from the ship's control room and can be executed automatically or manually. The dock can be flooded in 90 minutes to a depth of up to 3m and it is operable up to Sea State 4. If landing craft are not embarked, the dock can serve as a bay for additional tanks or other vehicles. The hangar/garage areas can be used for other payloads such as containerised accommodation, palletised stores, medical equipment or even as space for evacuated non-combatants.

Vehicle and cargo handling is carried out through a combination of lifts, cranes and access doors. There are two 19-ton elevators, one located forward of the superstructure on the starboard side and one located closer to the centreline aft. A 5-ton ammunition lift brings ordnance up to the flight deck from a deep magazine. For embarkation and offload alongside, the ship is fitted with a lateral starboard ramp door giving access to the main garage area. A 20-ton crane situated in front of the island structure can also be used to load and unload equipment from the quayside.

The designers have put great emphasis on the vessel's survivability. The ship is divided into several independent damage control (DC) zones separated by watertight bulkheads that incorporate fireproof insulation. Each zone features its own fire-fighting capabilities, ventilation systems, air-conditioning modules, electrical power distribution panels and chemical (NBC) filtration systems. Damage control co-ordination is supported by an integrated battle-damage monitoring and control system (IBDCS) centred in the machinery control room and widely accessible from a network of combat consoles distributed onboard. A large number of automated detectors, sensors and cameras linked to the ship's

A truck being lowered by *Dokdo*'s forward elevator, which is located in front of the forward island. The hangar and garage areas can be accessed by means of the two aircraft/vehicle lifts. *(Republic of Korea Navy)*

CCTV system are fitted throughout the ship, ensuring early warning of any damage that might occur. The system records, processes and reports specific information, thus enabling accurate identification of the ship's platform status at all times. The ship is built to support operations without any significant damage up to Sea State 7 and with flight operations possible up to Sea State 5 by using a fin stabiliser system.

PLATFORM SYSTEMS

The functionality of the damage-control system mentioned above is achieved by means of an integrated platform management system (IPMS) supplied by Canadian company CAE marine controls division (now part of L-3 MAPPS). It has been based on experience gained with similar systems developed, for example, for Germany's *Sachsen* class and the Royal Netherlands Navy's *De Zeven Provinciën* class air-defence frigates. The comprehensive system monitors and controls the ship's hull, propulsion, electrical and mechanical devices, thus ensuring permanent status identification and control of all internal technical components. The IPMS is equipped with a comprehensive range of user interfaces, with its various subsystems connected to a dual-redundant main internet data bus. There is provision for interface and information exchange with both the integrated bridge system (IBS) and the combat management system.

The integrated bridge system was supplied by L-3 MAPPS/Samsung. It includes all navigation sensors, an automated identification systems (AIS) to help locate and identify other ships, and a warship electronic chart display and information system (WECDIS).

COMMAND AND CONTROL

The ship is equipped with a Samsung Thales Corporation combat management system (CMS), which is based on the popular Thales Nederland TACTICOS series. It benefits from a fully modular hardware and software, providing the ship with digital C4ISTAR (command, control, intelligence, surveillance, targeting, acquisition and reconnaissance functions).

From a hardware perspective, the CMS is comprised of double- and triple-screen high resolution LCD multi-function operator consoles, incorporating a keyboard with roller-ball and joystick.

A starboard view of *Dokdo*, showing both the side vehicle ramp and 20-ton crane. These can be used for loading and unloading purposes when a quayside is available. *(Guy Toremans)*

There are also two large-screen tactical displays. The CMS's concept is based on providing the operator with all the information, data and controls necessary to perform allocated operational tasks. All consoles and workstations are connected by means of a redundant fibre-optic data transmission system, which also handles signals from the IPMS

Three starboard views of *Dokdo*'s island superstructure showing the layout of key sensors and other equipment. A bulbous Selex Galileo SPN-720 precision approach array is mounted on a platform to the rear of the island, with a Goalkeeper CIWS located on the deck aft of the second funnel. The main SMART-L search radar is mounted on top of the mainmast between the two funnels, whilst a Dagaie Mk 2 launcher protrudes from a platform just forward of this mast. A TACAN beacon can be seen to the top rear of the forward mast, with cylindrical 'pots' for the ESM components of the SLQ-200(V) Sonata system located at the head of and on platforms either side of this mast. The MW-08 target-indication radar is mounted high on the forward part of the mast, with the SPS-95K search and navigation radar on a platform somewhat further down. The starboard jammer that forms a key part of the Sonata system's ECM capabilities is at the base of the foremast, with the Mk 49 launcher for the RAM system above the bridge. *(Guy Toremans)*

The CMS provides situational awareness across the air, land, surface and sub-surface environments by assimilating and associating multi-source information from both organic (shipboard) sensors and non-organic (external) information feeds. For example, Link 11 and Link 16 tactical data links allow interface with other national and allied units. The system controls and manages the ship's weapons and processes, as well as exchanging tactical information with the combat systems of other units. It also supports the planning and decision-making process, not least through its capacity to generate and sustain a recognised maritime picture. All this makes the ship well suited to act as a command platform for joint and amphibious operations.

SENSORS: THE SHIP'S EYES AND EARS

Dokdo's sensor suite makes an interesting comparison with similar European vessels such as the *Mistral* and *Juan Carlos I* classes, being comparatively more elaborate. It includes the sophisticated Thales SMART-L 3D volume search radar that is found on many 'high end' air-defence classes, as well as a Thales MW-08 surface/target-indication radar. Other systems installed include a GECO Engineering SPS-95K surface-search/navigation radar and a Selex Galileo SPN-720 precision approach-control radar.

Dokdo's principal defensive armament is a RAM Mk 31 Mod 1 Guided Missile Weapon System (GMWS), which includes a 21-cell Mk 49 launching system and RIM-116 Rolling Airframe Missiles. This picture shows the launcher on top of *Dokdo's* bridge. Two Goalkeeper CIWS also provide last-ditch defence. *(Guy Toremans)*

Thales SMART-L Volume-Search Radar: Designed to provide long-range air/surface surveillance and target designation the SMART-L radar permits the early detection and tracking of even very small aircraft and missiles. Operating in the NATO 1,000–2,000 MHz D band (US Navy L Band), it provides very long-range coverage, reportedly being able to detect a low observable target at 35nm (65km), a fighter at 120nm (220km), and a maritime patrol aircraft at over 215nm (400km). Its tracking file can handle up to 1,000 aircraft or 100 surface targets simultaneously. The antenna is electronically stabilised against ship movement, up to the extent of a 30° roll and a 15° pitch. The information SMART-L provides makes an essential contribution to the threat evaluation process – especially in multiple attack scenarios – and allows the weapon control system to lock-on rapidly to potential targets.[5]

Thales MW-08 Target Indication Radar: Operating in the NATO 4,000–6,000 MHz G band (US Navy C band), MW-08 is used for surveillance and target indication at ranges out to c. 80nm (150km) and can also be used to provide gun fire control.

GECO Engineering SPS-95K Search/Navigation Radar: This is composed of a transceiver, a remote control unit and a reflector with omni IFF antenna. It has an instrumental range of c. 107nm (200km).

Selex Galileo SPN-720 Shipborne Precision Approach Radar: This is designed to assist the rapid take-off and recovery of aircraft in both day and night-time conditions, including during adverse weather. The radar system is able to control two aircraft simultaneously and provides final approach and deck landing guidance. The system employs a Doppler radar that uses monopulse tracking out to a range of 15nm (28km). SPN-720 can be operated as a stand-alone system or integrated with a ship's combat management system.

Other sensors fitted include a Thales TACAN tactical air navigation system and the Sagem Vampir (*Veille Air Mer Panoramique Infra Rouge*) IR tracking system. The latter provides 360° passive panoramic surveillance coverage through a high-performance camera system, detecting, tracking and reporting threats that could range from sea-skimming anti-ship missiles through to fast incoming attack craft or even jet-skis. Detection ranges, dependent on the target, can extend to around 15nm (27.5km) or more.

In addition to its powerful suite of sensors, *Dokdo* also benefits from effective electronic warfare and physical countermeasure capabilities. Although details are somewhat sketchy, it appears that both electronic countermeasures (ECM) and electronic support measures (ESM) are provided by an indigenous Nex I Future SLQ-200(V) Sonata system, which is similar to that installed in a number of South Korean surface escorts. A Korean SLQ-260X TACM (torpedo acoustic countermeasures) torpedo-defence system can also be used to launch decoys to deflect incoming torpedoes.[6] There are also two French Dagaie Mk 2 chaff launchers.

COMMUNICATIONS SUITE

An integrated communications suite (ICS), delivered by UNIMO Technology, provides the necessary means for the digital management of both the internal and external onboard communications. The system operates across the full radio frequency spectrum and provides Link 11, Link 16 and civil and military satellite communications. It supports a message processing system, high speed data transmission, and a maritime tactical wide-area network.

WEAPONS SYSTEMS

Given the nature of her envisaged roles, *Dokdo* mounts only limited self-defence capabilities, as she would inevitably sail in consort with supporting escorts in higher-threat scenarios.

The ship's weapon suite includes a 21-cell Raytheon Mk.49 launch system for the RIM-116B Block I Rolling Airframe Missile (RAM). The Block 1 system configuration features upgraded infrared homing surface-to-air missiles that carry a blast fragmentation warhead. They travel at a speed of Mach 2 and have a reported range of nearly 5nm (9km). The system is primarily intended to provide close-point defence against incoming missiles. The RAM is scheduled to be replaced by the Korean Surface-to-Air Anti-Missile (K-SAAM) system in 2018. Housed in a four-cell VLS the 2.07m-long K-SAAM employs inertial mid-course guidance and an infrared seeker for terminal guidance.

A picture of *Dokdo* at speed during sea trials in February 2007. A CODAD propulsion system provides speeds of up to 23 knots, appropriate for the ship's amphibious role but less suitable for mobile task force operations. *(Republic of Korea Navy)*

Dokdo also mounts two seven-barrelled Thales Goalkeeper 30mm Close-In-Weapon Systems (CIWS), capable of firing 4,200 rounds a minute out to a maximum range of 2,000m. The Goalkeeper is a fully automatic system with a fast 'switch-over to the next-priority' target, assuring the timely detection and engagement of small and supersonic targets, even in dense clutter and jamming environments.

Although not featuring any inherent anti-surface warfare or anti-submarine warfare capabilities, specialised sea-control helicopters could be embarked to provide these capabilities.

Dokdo practicing an amphibious assault during a ceremony in September 2015 to commemorate the 65th anniversary of the landings at Incheon during the Korean War. The image also shows a LSF-II hovercraft, the ship's main waterborne assault vehicle, and other amphibious vessels used to prosecute Dokdo's primary amphibious role. *(US Navy)*

PROPULSION SYSTEM

The ship's propulsion system is based on a combined diesel and diesel (CODAD) plant, made up of four Doosan Heavy Industries/MAN Diesel SA 16 PC2.5 STC medium-speed engines. These can deliver up to 41,615shp (30.6MW). The engines, equipped with a sequential turbo charging (STC) system, are fitted on 'resilient' mountings and drive two shafts fitted with two variable-pitch propellers. The four engines are divided into two paired groups, separated by two auxiliary compartments, to improve survivability. The propulsion plant provides *Dokdo* with a maximum speed of c. 23 knots and an operational range of 11,000nm at 15 knots. Due to her high superstructure, wind conditions can have a significant effect on *Dokdo*'s manoeuvrability. A bow thruster has therefore been fitted to compensate for this effect and to enhance control during low speeds. The main propulsion system and auxiliary diesel generators and associated switchboards are monitored from the machinery control room/ship control centre by means of the power management and surveillance functionality of the IPMS.

CAPABILITY AND MISSION PROFILE

Without any doubt, *Dokdo* gives the South Korean Navy an important quantitative and qualitative leap in capabilities. The ship's advanced logistic features, the presence of the well deck and flight deck, coupled with extensive medical facilities, give her great flexibility of employment in both national and multinational environments. *Dokdo* is symbolic of a profound shift in ROKN thinking towards conducting expeditionary operations at extended range and endurance. It is an example of a technologically advanced, multi-purpose maritime platform able to operate in a joint/combined environment and adapt to any one of her four assigned mission profiles.

Amphibious Capabilities: *Dokdo*'s most obvious – and primary – task is to participate in amphibious operations, essentially the transportation, disembarkation and recovery of the 720-strong marine battalion and supporting equipment and provisions she is able to embark. The ship is able to support this capability for up to 45 days and can embark and support additional troops, notably by setting-up billet containers in the hangar/garage.

As for all amphibious assault ships, deployment of the embarked force can be carried out by airborne assault utilising troop-carrying helicopters operating from the flight deck or by means of landing craft carried in the well deck. The Republic of Korea Armed Forces operate both UH-60 Black Hawk and UH-1H Iroquois utility transport helicopters and the domestic KAI KUH-1 Surion is under development. Although indigenous LSF-II LCACs are the type most commonly operated from the well deck, the dock can accommodate a variety of landing craft types, as well as AAV-7 amphibious assault vehicles.[7]

Aviation Support Capabilities: In addition to being able to operate utility transport helicopters in the amphibious assault role, *Dokdo*'s aviation facilities are sufficiently flexible to support a wide range of aviation assets. Nominal helicopter capacity is commonly quoted as being ten helicopters, although this would inevitably be dependent on helicopter

A US Navy LCAC undertaking trials of *Dokdo*'s well deck soon after the ship was first commissioned in 2007. The LCAC's high speed allows supporting ships to be located further from an assault area than might otherwise be the case. *(US Navy)*

An image of a Korean LSF II LCAC. Built indigenously by Hanjin Heavy Industries, the design clearly owes much to similar US Navy craft. *(Guy Toremans)*

size and also other equipment embarked. The strengthened flight deck is capable of handling heavy types, up to and including the United States' MV-22B Osprey. This capacity was proven on 26 March 2015 when two US Marine Corps aircraft from *Bonhomme Richard* (LHD-6) performed touch-and-go landings onboard *Dokdo* for the first time.

Although *Dokdo* has the theoretical capacity to operate V/STOL aircraft this has never been practically demonstrated. The lack of a ski-jump – installation of which seems to have been considered – would, in any event, have a significant impact on take-off with a useful weapons payload.

Command Platform/Flagship: *Dokdo* possess an embarked force command support infrastructure, comprising extensive command, control and communications (C3) capabilities, anddedicated staff planning facilities, as well as office spaces. All of these have direct access to the ship's combat management centre. These facilities would provide working space for a command staff of up to 150 people, which could be increased to 200 for surge roles. The facilities are sufficiently flexible to allow the ship to

One of *Dokdo*'s most important features is her full-length flight deck, which provides five operating spots. This pictures shows Korean UH-1H Iroquois and UH-60 Black Hawk helicopters operating from the four forward spots. The fifth spot has greater clearances and can support operation by a medium helicopter. The two elevators, both offset to starboard, can also be seen. *(US Navy)*

Dokdo seen operating with the US Navy *Nimitz* class aircraft carrier *George Washington* (CVN-73) in July 2010. Although much smaller than the US Navy ship, *Dokdo* is equipped with considerable command and control facilities that would allow here to lead an amphibious operation or fleet deployment. *(US Navy)*

control a three-dimensional joint amphibious operation or a fleet deployment, for example of a strategic mobile fleet division comprising KDX-11 and Aegis-equipped KDX-III destroyers and submarines.

Humanitarian Relief Vessel: When dealing with non-combatant operations such as disaster relief or humanitarian aid, the ship can rely on extensive medical facilities, giving her the capability to take up

Large numbers of members of the public can be seen on *Dokdo*'s flight deck as she sails in company with the KDX-II type destroyer *Choe Yeong* during an event to mark the anniversary of the Korean War in October 2015. The image demonstrates something of *Dokdo*'s capacity for humanitarian relief and civilian rescue operations, an important secondary role. *(Republic of Korea Navy)*

A Korean K1 series tank in *Dokdo*'s well deck. Note the sheathing on the dock's walls, which is intended to limit damage from landing craft berthing operations. . As well as conducting amphibious operations, *Dokdo* is sufficiently flexible to be used to transport large quantities or army and air force equipment. *(Guy Toremans)*

A view of *Dokdo*'s island showing the two-level bridge – the lower being a staff bridge – and the flight-control centre. The second vessel in the class – *Marado* – is likely to expand the aviation capabilities inherent in the *Dokdo* design. *(Guy Toremans)*

the role of a primary casualty receiving ship. These medical facilities include a fully-equipped hospital area incorporating two operating theatres, a triage area, an X-ray room, sterilisation area, laboratory, an isolation room, an intensive care unit (ICU), a post-operative room, a quarantine cabin, a pharmacy and a consulting room. There are good connections to the flight deck, garage area and well deck. During normal operations *Dokdo* typically carries two doctors and four nursing officers. However, when deployed in support of humanitarian missions, this medical team would be augmented by additional personnel. If necessary modular field hospital units can also be taken onboard and stored in the garage areas.

Dokdo is also sufficiently flexible to undertake a variety of other roles, include transportation of army or air force equipment, the provision of support to a mine-countermeasures operation or deployment as a training ship.[8]

AFTER *DOKDO*: THE WAY AHEAD

The commissioning of *Dokdo* marked only the first stage in the recapitalisation of South Korea's amphibious forces and the process is still ongoing. Certainly, events of the last few years have demonstrated the need for mobile and versatile forces that are able to react quickly to crisis and conflict scenarios. This is leading to the ROKN to look at more advanced amphibious platforms that can be used to meet both local security requirements and expeditionary ambitions.[9]

A major development has been confirmation of the long-awaited order for a second member of the class, *Marado*, again from Hanjin Heavy Industries & Construction, in December 2014. Launch is expected within the next twelve months and it is likely that the new ship will be fully operational by 2020. The press release announcing the contract for the second ship suggested that the design would upgrade the capabilities deployed by the current *Dokdo* '… through the adoption of the latest technology and advanced equipment'. There has been considerable speculation as to what this may mean in practice. Much of this has inevitably featured on *Marado*'s aviation facilities, including potential for V/STOL operations through incorporation of a ski-jump and other flight-deck modifications. The ability to support fast jet operations would provide the ROKN with significant additional flexibility to deal with low- and medium-intensity combat operations. It is also noteworthy that the land-based F-35A variant of the Lockheed Martin Lightning II Joint Strike Fighter has also been selected for the air force's requirements, facilitating a F-35B acquisition. Other speculation surrounding *Marado* has concerned the ship's size – she may be somewhat longer and have a higher displacement to support a larger air group – and propulsion system. For example, the selection of a high-powered CODOG (combined diesel or gas) of CODAG (combined diesel and gas) would provide a higher maximum speed and, thus, greater synergies with the surface vessels forming the strategic mobile task groups.

The ROKN retains a long-standing requirement for a third aviation support ship. Some sources suggest that this ship will be ordered before the end of the decade and commission before 2025. The name *Baengnyeongdo* has already been provisionally allocated to the ship.[10] If constructed, it would seem likely that the new ship would be built to an enlarged design and a significant step on the way towards an ambition to field a true aircraft carrier capability.

A MV-22 Osprey from the US Marine Corps pictured operating with *Dokdo* on 26 March 2015, the first time the type had landed on the ship. *Dokdo* benefits from considerable flexibility in terms of the range of rotary-wing aircraft she can deploy but she has yet to embark fast jets. However, there has been speculation that the flight deck of *Marado*, the second member of the class that is currently under construction, will be modified to support F-35B operations *(US Marine Corps)*

Notes

1. See ROKN Headquarters' *Navy Vision 2020* (Gyeryongdae: Republic of Korea Navy, 1999). The document has subsequently been refreshed by *Navy Vision* 2030, published in 2008.

2. Published reports on the LPX programme have often been rather sketchy and sometimes contradictory, with some sources suggesting a requirement for a fourth ship. However, the strategic mobile fleet construct developed by Admiral Ahn Byoung-Tae was based on three task groups and this seems to have been reflected in the number of ships formally planned. Further insight into the ROKN's thinking at this time is contained in Ian Bowers' thesis, *The Creation of the Modern Republic of Korea Navy* (London: King' College London, 2013) which can currently be downloaded by searching: https://kclpure.kcl.ac.uk/portal/

3. Dokdo, also known as the Liancourt Rocks or as Takeshima in Japanese, comprises a group of small islets in the Sea of Japan. Sovereignty of the group is disputed between South Korea, which controls the islets, and Japan. The use of the name on the ROKN's largest ship therefore carries considerable national significance.

4. Two South Korean-designed LSF-II hovercraft manufactured by Hanjin Heavy Industries are normally embarked. Displacing around 150 tons in full load condition, they are 26.8m long and 14.3m wide. Capable of carrying up to twenty-three troops or 55 tons of vehicles and equipment at speeds of up to 40 knots in Sea State 2, the craft bear a strong resemblance to the US Navy's LCAC.

5. It is worth noting that when *Dokdo* was launched, a wooden mock-up version of the SMART-L radar system had been installed because the actual system was still being tested at Thales Nederland's headquarters in Hengelo.

6. Although the installation of TACM is mentioned in several sources, there is no conclusive evidence the system has actually been installed on the ship.

7. The AAV-7 is a fully-tracked amphibious vehicle that provides an over-the-beach capability for up to eighteen troops and their equipment. It is the only vehicle in the world capable of operating in rough seas and plunging surf conditions.

8. Although *Dokdo* is undoubtedly a capable and flexible ship, there has been some criticism of the design, perhaps reflected in the amount of time between her delivery and the order for a second ship of the type. One particular problem appears to have been a limited maximum speed in the order of 23 knots. This has restricted her value in the planned role as flagship of one of the strategic mobile groups, which otherwise include faster surface escorts of the KDX-II and KDX-III types. It is interesting to note that she has been assigned to amphibious squadron rather than Maritime Task Flotilla Seven structure established to further the strategic mobile concept. One commentator has even gone so far as to say, 'Up to this point, the ship has been a failure and until a more concrete role has been found the ROKN will struggle to utilise it effectively.' – Ian Bowers, *The Creation of the Modern Republic of Korea Navy*, op. cit., p.142.

9. In addition to amphibious assault ship types represented by *Dokdo*, the ROKN is also constructing a new class of *Cheon Wang Bong* tank landing ships. At nearly 5,000 tons standard displacement, these ships resemble small amphibious transport docks and are intended for operations in both national waters as well as overseas. The first of the class entered service in December 2014 and a total of four are planned.

10. Continuing a trend of controversial names, *Baengnyeongdo* will be named after Baengnyeongdo Island, which is the most remote part of the South Korean territory facing North Korea across the Northern Limit Line – the de facto maritime boundary between the two countries – in the Yellow Sea. It is the spot closest to the 38th Parallel on the southern side and the location of significant combat, both during the Korean War and more recently. By way of contrast, Marado Island is not a disputed territory, simply representing the southernmost point of Korean territory.

1000

3.3 SIGNIFICANT SHIPS

Authors:
Edward Feege & Scott Truver

ZUMWALT (DDG-1000)

PAST AND FUTURE TENSE

As America's newest destroyer *Zumwalt* (DDG-1000) made her way towards the Atlantic Ocean and her initial sea trials in December 2015, observers on the banks of Maine's Kennebec River could be forgiven for wondering if they were gazing upon a relic of a past era – or the harbinger of things to come.

A cruiser in all but name, the 15,800-ton *Zumwalt* 'land-attack destroyer' was conceived in the 1990s, when the US Navy's greatest challenge was to operate and fight in the world's near-shore, littoral environs. That primarily translated into the projection of power ashore, and supporting operations against 'rogue' nations. This was a tall order, and one that assumed that the United States could concentrate more on the plethora of shorter-range 'green-water' threats closer to shore than on 'blue water' threats represented by the Soviet navy that had imploded with the fall of the Berlin Wall.

That situation changed during the next decade. As if overnight, the US Navy was faced with a myriad of 'anti-access/area-denial' (A2/AD) challenges from rising – or recovering – powers such as China and Russia and from regional powers and non-state actors that were gradually improving their arsenals. Increasingly, assured access by sea to key areas of the world was no longer certain. Given this shift, and growing programme costs, the DDG-1000 class no longer looked as critical, or as affordable, as originally envisaged. The US Navy's enthusiasm for these ships waned, and Service and

Left: The lead *Zumwalt* (DDG-1000) class destroyer pictured at speed in the Atlantic Ocean whilst undertaking acceptance trials on 21 April 2016. A cruiser-sized vessel of radical appearance, *Zumwalt* incorporates numerous technical innovations that are likely to have a significant impact on future US Navy surface combatants. *(General Dynamics Bath Iron Works)*

The *Zumwalt* class traces its origins to 1990s plans to supplement existing multi-mission *Arleigh Burke* (DDG-51) class destroyers and *Ticonderoga* (CG-47) class cruisers with a cheaper single-mission combatant – the DD-21 – focused on littoral warfare. Although the DD-21 design was ultimately scaled back, the new warship is still considerably larger and more expensive than the ships that were intended to form the 'high end' of this force mix. As a result, orders were reduced to just three ships in favour of renewed production of the DDG-51 type; *Donald Cook* (DDG-75) is shown here. *(US Navy)*

The CG-47 class cruiser *Monterey* (CG-61). Continued investment in 'high end' multi-mission combatants such as the CG-47 design was the US Navy's preferred option at the end of the Cold War but budget constraints resulted in a two-tier force structure plan. The proposed DD-21 design – from which *Zumwalt* emerged – formed the 'low end' of this mix. *(US Navy)*

Department of Defense officials ultimately decided to limit the number of ships in the class to three, down from thirty-two when the requirement for a new destroyer was first conceived.

The fact that *Zumwalt* and her two sisters survived at all was testament to the promise that the warships held for the future US Navy, not the littoral-focused service of the 1990s. The ships in the class have numerous features that push the state of the art in American naval combatants. But the key to DDG-1000's survival, and possibly the US Navy's future, lay in only a handful of areas, particularly shipboard electrical power.[1]

PROGRAMME ORIGINS

The *Zumwalt* class had its genesis in the aftermath of the victory of the United States and its coalition partners over Iraq during Operation 'Desert Storm' and the first Gulf War. In 1992, then-Chief of Naval Operations Admiral Frank Kelso commissioned the '21st Century Destroyer Technology' Study to establish the requirements for a new surface combatant to be built in the mid-to-late 2000s. The study was influenced by several aspects of the post-'Desert Storm' world, including the lack of a well-defined threat and a corresponding framework for American operations. Fiscal constraints, that had been growing more stringent since the end of the Cold War in 1989, also loomed large, as did the navy's experience in the war with Iraq and its overwhelming focus on projecting power ashore.[2]

In a nod to the fiscal environment the navy was facing, the study participants concluded that the preferred option – a surface combatant force consisting solely of high-end, multi-mission warships – would not be realistic. Instead, they recommended a two-tier 'family' of ships, with multi-mission warships comprising the 'high' end of the force, while other combatants – focused on one particular mission – made up the 'low' end. The requirements for this Surface Combatant (SC)-21 family of ships concept was blessed by the Joint Requirements Oversight Council – the Vice Chairman of the Joint Chiefs of Staff, the second highest-ranking US military officer, and the four chiefs of the military services – in June 1994. The navy subsequently performed a formal analysis of alternatives, which established the force architec-

A graphic of the approved *Zumwalt* design, dating from around the time funding for the first two ships was approved under the FY2007 budget. The design featured fewer missiles and a lower magazine capacity than the initial DD-21 project but otherwise kept many of its predecessor's features. *(US Navy)*

ture it would require to meet its war-fighting requirements, and the performance it demanded of the ships within the force. Planners determined that a focused-mission, land-attack combatant would be the first of the SC-21 family of ships to be ordered.[3]

Rear Admiral Daniel J Murphy, the navy's Director of Surface Warfare, laid out the key parameters for this all-important power-projection mission in an article in the US Naval Institute's *Proceedings* journal in June 1997, viz.

- Surface fire (gun) support for US forces ashore at ranges of up to 75 miles (120km) inland.
- Interdiction fire to break up armour and troop concentrations at ranges of 200 miles (322km) inland.
- Deep strikes against enemy command-and-control nodes, air defences, and ballistic missile forces at ranges of 1,000 miles (1,600km) or more.[4]

US Navy efforts to meet these ambitious requirements led to concepts for two significantly different combatants: A Maritime Fire Support Ship and the DD-21 land-attack destroyer. The former concept – an austere missile carrier, equipped with 512 vertical launching system (VLS) missile tubes and manned by a small crew – was also known as the 'arsenal ship'. The arsenal ship concept was first relegated to the status of technology demonstrator and then ultimately cancelled in 1997.[5]

That left DD-21, a warship able to help achieve maritime dominance but with a focus on land attack. Its capabilities in the former mission area – which included anti-submarine and anti-surface ship warfare – would allow the new class of thirty-two DD-21s to replace the aging *Spruance* (DD-963) class destroyers and *Oliver Hazard Perry* (FFG-7)

The proposed DD-21 design featured a number of characteristics intended to help it fight and survive in a challenging littoral environment, including a drastically reduced radar cross-section achieved by use of a tumblehome, wave-piercing hull design. This feature survived into the recast DD-(X)/DDG-1000 design and is one of a range of technologies tested in the Advanced Electric Ship Demonstrator (AESD) *Sea Jet*. A 41m, roughly quarter-scale model of an actual destroyer, *Sea Jet* is operated by the Naval Surface Warship Center Carderock MD from its acoustic research detachment at Bayview ID. *(US Navy)*

Two views of *Zumwalt* under construction at Bath Iron Works (BIW) in Maine on 15 October 2013, shortly before float out. The ship rests on a wheeled transfer system that allowed movement into the floating dry dock used for launch. The middle section of the second ship in the class, *Michael Monsoor*, is positioned astern. The images clearly show the tumble-home, wave-piercing hull design – used to reduce radar cross-section and therefore improve survivability in a littoral environment – that has survived from the original DD-21 land-attack destroyer concept. *(General Dynamics Bath Iron Works)*

class frigates. Its capabilities in the land-attack arena would allow it to close a gap in the ability to provide surface fire support to US Marine Corps and other forces ashore. That shortfall had worsened when the Navy retired its last two Second World War-era *Iowa* class battleships after 'Desert Storm'.[6]

Simply put, the DD-21s would be able to fight and survive in the challenging littoral environment and deliver high-volume gun and missile fires in support of ground and air campaigns ashore. Some of the key ship characteristics that would facilitate this capability would include:

- A 1,400-round, twin 155mm vertical launch gun system (referred to as a VGAS, which later evolved into the Advanced Gun System, or AGS).
- Two 64-cell VLS packs 'capable of launching any missile in the 21st Century surface navy inventory'.[7]
- A drastically-reduced radar cross-section (RCS), achieved in part through the use of a tumblehome, wave-piercing hull design that scatters and diffuses radar reflections away from enemy receivers. This would be combined with a significant reduction in the ship's acoustic signature, which the navy announced would be achieved in large part through the use of quiet, electric-drive propulsion.
- A new radar geared towards ship self-defence, and a sonar optimised for underwater detection of submarines and mines in the littoral environment.
- A crew size of 95–150, smaller than that of any US Navy surface combatant then in service.[8]

And as Rear Admiral Murphy noted, the intent was to do all of this cost-effectively – the unit cost of the fifth ship in the class was not to exceed US$750 million in FY1996 dollars.[9]

In 1998, the US Navy, in consultation with its private-sector vendors, devised a plan under which two industry groups would compete to design and provide life-cycle support to the ships in the DD-21 class. The 'Blue' team was centred on General Dynamics and its Bath Iron Works (BIW) shipbuilding subsidiary, along with Lockheed Martin as the team's system integrator. Similarly, the 'Gold' team included Northrop Grumman and its Ingalls Shipbuilding subsidiary, paired with Raytheon, which would be that group's system integrator. Ultimately, the navy expected that both teams would split the actual construction of the planned destroyers.

A picture of *Zumwalt* being floated out on 28 October 2013. Note the raised 155m gun barrel of the forward AGS mounting. The two AGS mountings, which provide the class's main land-attack capability, have been a significant driver of overall ship size. *(General Dynamics Bath Iron Works)*

As the programme moved forward, DD-21 also acquired a name. In July 2000, the US Navy announced that the first ship in the class would be named for Admiral Elmo R Zumwalt, the Chief of Naval Operations from 1970 to 1974. But even at this stage, however, events were conspiring to drastically reduce the class's size and budgetary impact.

THE GREAT TRUNCATION

As preliminary designs for *Zumwalt* moved forward, the US Navy realised that the requirements for its new destroyer class were driving up the ship's size and costs. DD-21's estimated full-load displacement grew from an initial 9,000 tons to between 16,000 and 18,000 tons. This was driven in part by a combination of the low-RCS tumblehome hull and the outfit of guns, VLS and missiles the ship would carry. In comparison, the Navy's *Ticonderoga* (CG-47) class cruisers had a full-load displacement of only 9,600 tons. That said, inside-the-lifelines volume was cheap; it was what was going to be installed that soon generated 'sticker shock'.

With time, the navy gained a better appreciation of the technical risks associated with the development of DD-21's electronics, weapon systems and propulsion. Consequently, the US$750 million price tag for a fifth ship in the class looked increasingly unattainable. Cost growth in ship programmes is not unusual, but the potential magnitude of that growth – and the impact burgeoning DD-21 costs might have on other US Navy programmes – increasingly pointed to the need for a course correction.

That came in November 2001, when the Navy announced that the DD-21 had been replaced by a new DD(X) Future Surface Combatant Program. This programme included a 'family' of ships, including a DD(X) destroyer, a multi-mission CG(X) cruiser to handle air and missile defence, and a smaller littoral combat ship (LCS) to deal with submarine, mine and fast attack craft threats closer to shore.[10]

DD(X) essentially would be DD-21 – but smaller. DD(X)'s armament, sensors, low-RCS features, crew size, advanced automation, electric drive and other features would be similar to its predecessor. But in a nod to affordability, the number of VLS cells would shrink from a planned 128 on the DD-21 to eighty on DD(X), while magazine capacity for DD(X)'s AGS mounts would also be reduced. However, DD(X) would add new dual-band radars – an X Band (NATO I/J Band) multi-function radar and an S-Band (NATO E/F Band) volume-search radar – giving ships in the class an area air-defence capability not originally planned.

US Navy planners believed that those changes would bring the new ship's displacement below 15,000 tons and decrease costs. Nevertheless, DD(X) – still named *Zumwalt* – would be the largest surface combatant to serve in the US Navy since the nuclear-powered cruiser *Long Beach* (CGN-9) of 1961. Moreover, the estimated cost of the fifth ship would remain over US$1 billion. These stubbornly high programme costs led the navy to reduce its acquisition plans to twenty-four ships, a fall of eight from DD-21. And that was not the end of the shrinking horizons for DD(X). Between 2001 and 2005, the number of ships in the programme was cut again, first to between eight and eleven, and then to between five and six ships.

Aware that these costs were still excessive for a

Table 3.3.1.

ZUMWALT (DDG-1000) PRINCIPAL PARTICULARS

Building Information:	
Fabrication Commenced:	11 February 2009[1]
Launched:	28 October 2013
Commissioned:	15 October 2016 (planned)
Builders:	General Dynamics Corporation's Bath Iron Works yard, Bath ME, USA.
Dimensions:	
Displacement:	15,761 tons full load displacement.[2]
Overall Hull Dimensions:	186.0m x 24.6m x 8.4m (sonar).
Equipment:	
Missiles:	20 x Mk 57 4-cell VLS modules. Can accommodate most current and planned US Navy vertically-launched missiles. Will initially be equipped with BGM-109E Tomahawk land-attack cruise missiles and RIM-162 ESSM.
Guns:	2 x 155mm Advanced Gun Systems (AGS). 2 x 30mm Mk 46 GWS (yet to be fitted).
Aircraft:	2 x MH-60R Seahawk helicopters or 1 x MH-60R and 3 x MQ-8 Fire Scout series UAVs.
Boats:	Internal bay for up to 2 x 11m RHIBs.
Countermeasures:	Unspecified ESM/ECM system. Space reserved for AN/SLQ-25 Nixie torpedo-defence system.
Principal Sensors:	AN/SPY-3 multi-function radar. Navigation radar. AN/SQQ-90 integrated undersea warfare suite comprising AN/SQQ-60 MF and AN/SQQ-61 HF bow arrays and AN/SQR-20 towed array.
Combat System:	Raytheon Total Ship Computing Environment (TSCE) linked to Common Display System (CDS) workstations.
Propulsion Systems:	
Machinery:	Integrated Electric Propulsion. 2 x Rolls-Royce MT30 gas turbines each rated at 35.4MW and 2 x Rolls-Royce RR4500 gas turbine generator sets each rated at 3.9MW produce a total of 78MW ship power. Propulsion through 2 x GE Power Conversion Advanced Induction Motors each rated at 33.6MW producing a total of c. 90,000hp through two fixed shafts.
Speed:	Designed maximum speed is in excess of 30 knots.
Other Details:	
Complement:	A typical crew comprises c. 147 personnel plus an additional 28-strong aviation detachment. Accommodation for 186 personnel.
Class:	Three ships have been ordered: *Zumwalt* (DDG-1000), *Michael Monsoor* (DDG-1001) and *Lyndon B. Johnson* (DDG-1002).

Notes:

1 Official keel laying was 17 November 2011.

2 There are significant variations in quoted displacement figures for the class. The quoted figure is from the US Navy's Naval Vessels Register as of June 2016.

ship focused on a single mission, the navy also reclassified DD(X) as a multi-mission combatant, now designated DDG-1000, in April 2006. Now six ships would follow the first ship in the class. The US Congress funded the first two ships of the class in 2007, with funding split across two fiscal years.

Even with this progress, further programme cuts were made in 2008. This time, however, the navy's leadership emphasised *Zumwalt*'s shortcomings in light of a changing strategic environment. Service officials announced that the class would be limited to three ships, and that the resources thus freed up would be devoted to the construction of additional *Arleigh Burke* (DDG-51) class destroyers, which formed the backbone of the service's surface warship force. As Chief of Naval Operations Admiral Gary Roughead explained, changes in the threat environment drove the Navy's decision:

The world has changed markedly since we began the march to DDG-1000 in the early 1990s. Between 1990 and 2006, for example, approximately one new nation developed a ballistic missile capability every three years and then in July of 2006 the terrorist group Hezbollah, not a nation, launched an anti-ship missile against the Israeli ship *Hanit*. DDG-1000 did not address these important changes

***Zumwalt* (DDG-1000)** (2015)
1:1000 scale

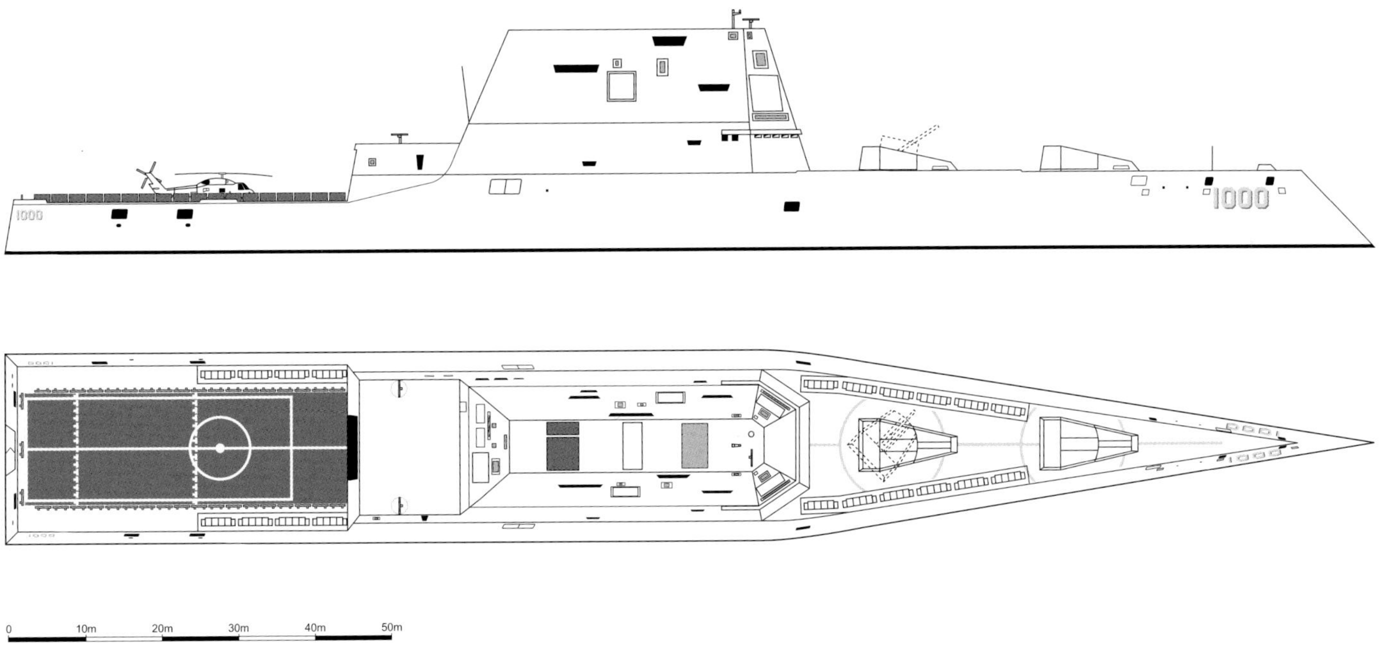

(Drawings © John Jordan, 2015)

An April 2016 view of *Zumwalt* on trials from General Dynamics' Bath Iron Works yard. Although a Northrop Grumman-led team won the initial detailed design contract for the class, all three members of the class have been built at the General Dynamics' facility. *(US Navy)*

> and challenges. In addition, while it has been optimised for littoral anti-submarine warfare, the growth of the worldwide submarine fleet – a growth projected by business sources to be 280 submarines over the next two decades – does not allow us to diminish our deep-water capabilities. For these reasons, I made the difficult decision to truncate DDG-1000, to take advantage of the technologies, to learn from them, but continue the DDG-51 line because it has the right capabilities and provides greater capacity where we need it.[11]

Cutting the class size meant that upfront development costs would be spread across three ships, not the thirty-two first envisioned. That in turn, caused an eighty-six per cent jump in unit cost, which in 2010 led to a breach under the Nunn-McCurdy provision of US acquisition law.[12] With the breach, the Under Secretary of Defense for Acquisition, Technology, and Logistics Ashton Carter (in 2015 confirmed as Secretary of Defense) stepped-in to determine the cause of the spike. Carter's office also worked with the US Navy to bring the programme back in line with acquisition law if possible, or terminate it if costs and performance could not be brought back under control.

The navy and the Office of the Secretary of Defense (OSD) agreed that the breach mainly was a function of cutting the size of the class. The programme was re-authorised to proceed, but with changes imposed. One was a design alteration – the ship would no longer feature a dual-band radar. The S-band volume-search radar was foregone, although the option to back-fit it at a later date was retained. That left the *Zumwalt* class ships with only an X-band multi-function radar, dubbed AN/SPY-3. With those decisions, the final structure of the class was in place.[13]

Names of the second and third ships were announced in 2008 and 2012. DDG-1001 eventually was named *Michael Monsoor* in honour of a Navy SEAL and Medal of Honor winner killed in Iraq in 2006. DDG-1002 was named after the 36th President of the United States, Lyndon B Johnson.

BUILDING THE *ZUMWALT* CLASS

In April 2002 the navy awarded the Northrop Grumman-led Gold team a US$2.88 billion contract to conduct preliminary design work on DD(X) and to build and test engineering development models of key systems to reduce the risks associated with the new technologies slated for the ship. The navy subsequently brought much of the Blue team into the fold in a bid to ensure that both Northrop Grumman's Ingalls Shipbuilding and General Dynamics' Bath Iron Works could both bid on the detailed design and construction of DD(X), beginning in FY2005.

In 2004, the navy and the OSD agreed on an acquisition strategy that would see Northrop Grumman's Ingalls yard in Mississippi build the first ship in the DDG-1000 class. General Dynamics' BIW would build the second ship at its yard in Maine. The remaining ships in the class – not yet truncated at three ships – would be split between both shipbuilders. The navy's leadership changed its mind in 2005, and asked for permission to hold a winner-take-all-competition between the two shipbuilding groups. But OSD reticence and the opposition of congressional delegations from Mississippi (home to Ingalls) and Maine (home to BIW) scuppered that approach. The navy then switched to a dual lead-ship approach, with each yard building one ship of the pair authorised in the FY2007 budget.

The navy's announcement limiting the class to three ships resulted in further changes. Three-way navy and industry talks took place in 2009, after which the service announced that GD/BIW would build all three *Zumwalt* class ships, while Ingalls would build a greater percentage of the additional DDG-51 destroyers also announced in 2008. Ingalls also would provide components for the DDG-1000s, including deckhouses made of composite materials. By that point, fabrication had already begun on *Zumwalt*.

CLASS CHARACTERISTICS

The mix of littoral combat-driven systems, combined with numerous state-of-the-art systems – some of which may be included in future warships – makes the DDG-1000 and its sisters a rare breed of warship. Some of the key attributes of these ships are discussed below. See also Table 3.3.1 for further technical details.

The distinctive deckhouses of the first two *Zumwalt* class ships are constructed of composite materials to reduce weight. They were the most significant of a number of components for the ships' provided by Ingalls shipbuilding business. The third ship, *Lyndon B. Johnson*, uses a more traditional steel structure. *(US Navy)*

Stealth and Survivability: One of the more striking aspects of the DDG-1000 design is the hull and superstructure, all shaped to lower the ship's RCS. Designers originally targeted a DD(X) RCS fifty times smaller than that of the *Arleigh Burke* (DDG-51) class ships. The decision to rely upon a tumblehome hull with a wave-piercing bow was a key part of this process. However, this type of hull shape also offered less internal volume than a more conventionally-shaped hull of the same size and displacement. Recovering internal volume that otherwise would have been lost – and adding reserve space and weight that could be used to adapt DDG-1000 for future systems and missions – led to the increase in the ship's size and displacement. At c. 15,800 tons, *Zumwalt* has a greater displacement than any other current US Navy surface combatant. The DDG-1000's tumblehome hull and wave-piercing bow might also create sea-keeping problems under more extreme conditions, as *Zumwalt*'s first prospective commanding officer, Captain James Kirk, noted:

> In extraordinary seas – wave heights greater than 20 feet, the ship can, if not sailed with care, experience undesirable righting motions. To address these concerns, extensive model testing at Naval Surface Warfare Center Carderock, Maryland, has been conducted, and a safe operating envelope across these higher sea states, which are expected for no more than 400 hours over the 35-year service life of the ship, has been developed.[14]

Indeed, during the December 2015 tests, which totalled 100 hours and ranged across the Gulf of Maine, US Navy and Bath Iron Works personnel tested the *Zumwalt* at full power – more than 30 knots – and ran the propulsion plant through more than a dozen different configurations. The tests also subjected the ship to a series of sharp turns (full rudder swings) to determine stability and overall manoeuvrability. This was to address concerns of observers who predicted that the ship would 'turn-turtle' during a sharp turn or in heavy seas. Naval architect Ken Brower told *Defense News* in 2007: 'The trouble is that as a ship pitches and heaves at sea, if you have tumblehome instead of flare, you have no righting energy to make the ship come back up. On the DDG-1000, with the waves coming at you from behind, when a ship pitches down, it can lose transverse stability as the stern comes out of the water—and basically roll over.'[15]

Those concerns have now seemingly been laid to rest. 'The faster it goes, the faster it responds', according to Rear Admiral (select) James Downey, *Zumwalt* programme manager, in January 2016 noting that the ship was able to stop in only 90 seconds.[16] The innovative tumblehome hull design had no issues with the eight- to ten-foot swells it encountered, and manoeuvrability was not affected at all.

DDG-1000's deckhouse is also shaped to reduce radar and infrared signatures. The first two ships of the class will rely upon deckhouses built with composite materials to effect this reduction and to remove weight. Their deckhouses are formed by panels and beams made of carbon fibre and vinyl ester skins, combined with balsa and foam cores. The ships' helicopter hangars, which are integrated into the overall superstructure, are formed with the same material.[17]

Many of the antennas for the ship's communication systems and sensors are embedded within their composite deckhouses, with the antenna faces flush with the deckhouse surface. However, as a cost-saving measure – taken in the aftermath of the 2010 Nunn-McCurdy cost target breach – the navy has apparently chosen to add a mast to the deckhouse design and relocated several antennae from the deckhouse to that mast, as well as to platforms on the side of the deckhouse. The move will increase the ships' RCS, although they still meet their overall signature-reduction targets.[18]

Northrop Grumman and subsequently its Huntington Ingalls Industries (HII) successor built the composite deckhouses at a facility in Gulfport, Mississippi, from where they were transported to BIW by barge.[19] That will not be required with DDG-1002, however, as the navy announced in 2013 that DDG-1002 would have a steel super-

This aerial view of *Zumwalt* gives a good impression of the ship's overall stealthy design, particularly the use of careful shaping of external surfaces to reduce radar cross-section. It is interesting to note the dividing line – just above bridge level – between the steel and composite deckhouse structure, as well as how the composite hangar has also been attached to the steel structure aft. Other details include the two stealth enclosures for the 155mm AGS mounts forward of the deckhouse. The launch cells for the Mk 57 VLS that forms the Peripheral Vertical Launching System are grouped around the edges of the deck. *(US Navy)*

structure instead of relying on a composite deckhouse. The service justified the move by noting that it had achieved greater weight savings with DDG-1002's design, and thus could forego the use of the composite deckhouse. The steel structure for the third ship is being built by BIW, which eliminates the transportation and integration costs associated with the previous arrangement.[20]

Weapon Systems: Given the *Zumwalt* class was designed around the naval surface fire-support mission, the new 155mm AGS is a central element in DDG-1000's arsenal. The early 1990s retirement of the US Navy's last battleships left only 5in/54-calibre (127mm) guns in the fleet to provide support to US Marine Corps and other troops ashore.[21] That system had a maximum range of only 13nm (24km), compared with the 23nm (43km) range of the battleships' 16in guns when firing unguided, unassisted weapons. At this short range, any warship providing gunfire support would be exposed to a range of anti-ship missile, rocket and artillery fire from shore defences. Only a longer-range weapon that allowed the firing ship to remain over the horizon from most enemy sensors and weapons would generate a reasonable chance of survival in a contested littoral environment.

For DDG-1000, the answer is the Mk 51 155mm AGS, an integrated gun and magazine weapon system capable of a maximum sustained firing rate of ten rounds per minute to ranges of up to 63nm (117km). The *Zumwalt* class's AGS are located in two bow turrets, each with its own automated magazine and weapon-handling system. The guns themselves incorporate thermal- and erosion-management technologies that help extend their barrel life and minimise infrared signature. Below deck, the magazines, which are unmanned, provide total storage for up to 600 rounds. And the control system for the AGS encompasses both gun and fire control, and is in turn integrated with the overall ship computing environment.

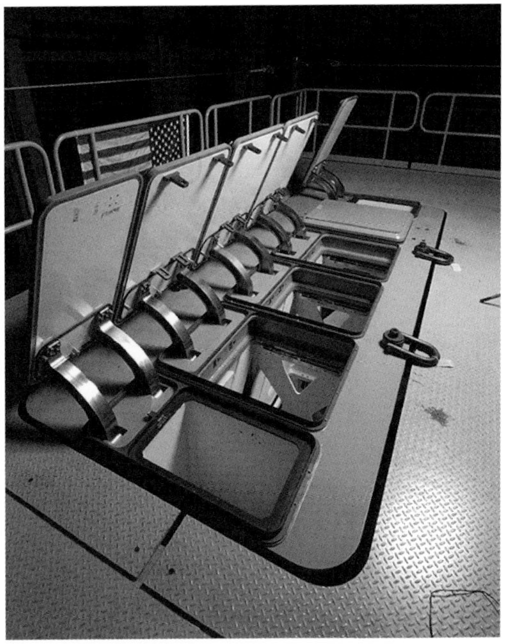

Two images showing a mock-up of a Mk 57 vertical launch system and its positioning on *Zumwalt* class ships. The Mk 57 system on *Zumwalt* comprises twenty four-cell units. Six of these are located on either side of the forecastle, with a further four on either side of the forward part of the helicopter deck. The system can handle most current and planned vertically-launched missiles in the US Navy's inventory. *(Raytheon, BAE Systems)*

A picture of a prototype 155mm Long-Range Land-Attack Projectile (LRLAP) undergoing testing in 2005. LRLAP rounds use rocket assistance and GPS guidance in conjunction with an inertial navigation system to provide extended range and improved accuracy for the *Zumwalt* class's AGS mounts. *(Northrop Grumman)*

The projectiles that the Mk 51 guns will fire are critical to the effectiveness of the AGS. The rocket-assisted 155mm Long-Range Land-Attack Projectile (LRLAP) features a unitary warhead, as well as hardened electronic systems capable of surviving the forces associated with its firing from a gun tube. These include a miniaturised Global Positioning System (GPS) receiver and inertial measurement unit receiver that provide guidance for the round. In addition to their range, these projectiles have a nearly vertical angle of fall during the final stage of their trajectory. This should help limit collateral damage when employed in urban or other crowded environments, according to LRLAP sub-contractor Lockheed Martin.

Missiles also provide an important offensive and defensive punch for the DDG-1000s. The *Zumwalt* class ships are equipped with the Mk 57 VLS, which consists of twenty, four-missile cells able to launch Tomahawk (BGM-109) cruise missiles, the SM-2, SM-3 and SM-6 variants of the Standard surface-to-air missile, and Evolved Sea Sparrow (ESSM, or RIM-162) self-defence missiles. Initially Tomahawk and ESSM will be installed. The systems can also support the Vertical Launch Anti-Submarine Rocket (ASROC), or VLA. In addition, the Mk 57 cells have room to accept future missiles that may be up to 25in (653mm) in diameter – greater than the 21in (533mm) tube limit of the Mk 41 VLS on other US Navy combatants. The Mk 57 uses an open software architecture, which likewise facilitates the addition of new weapons or system modifications, at a relatively low cost in time and resources.

Unlike other US Navy surface warships, the missile cells fitted in the *Zumwalt* class are arranged as a Peripheral Vertical Launching System (PVLS). This arrays the missile cells in rows at the edge of

the main deck. The inboard bulkheads of the system are heavily reinforced, so that any magazine explosion would be directed outward. There are twelve PVLS modules (six per side) on the forecastle of the ship, with eight flanking the flight deck aft (split four per side).

In addition to the major weapon systems mentioned above, the DDG-1000 class ships will also be fitted with two Mk 46 Gun Weapon Systems (GWS). Based on the Bushmaster 30mm chain gun – which fires up to 200 rounds per minute, to a maximum range of about 4,000m – the Mk 46 is designed to help the *Zumwalt* class counter swarming small-boat attacks in littoral regions. The US Navy substituted the Mk 46 for the 57mm Mk 110 gun system originally designated for use on board the new destroyers, with the service saying that the 30mm GWS was more cost-effective.[22]

Radar Systems: The Department of Defense decision to delete the class's planned S band volume-surveillance radar in response to the Nunn-McCurdy breach left the DDG-1000 class ships without a long-range surveillance and ballistic missile defence capability. In this regard, the *Zumwalt* class ships are inferior to the Navy's DDG-51 class destroyers and CG-47 class cruisers that are equipped with the AN/SPY-1 radar associated with the Aegis weapon system. The disparity with the in-work DDG-51 Flight III destroyers that will be fitted with the AN/SPY-6 air- and missile-defence radar (AMDR) will be greater still. However, the *Zumwalt* class retain the space and weight capacity to accept a long-range volume-search radar at some point in the future, should the Department of Defense and US Navy ever decide to reverse their position.

Nevertheless, as a modern multi-function active phased array, AN/SPY-3 still represents a considerable capability, particularly in the littoral environment. Operating in the X Band, the SPY-3 has a narrow beam width and wide frequency bandwidth, which are particularly useful for the detection and tracking of low-altitude and low-observ-

An external view of the hull-mounted elements of *Zumwalt*'s AN/SQQ-90 sonar suite. The sonar dome encloses an AN/SQQ-60 mid-frequency sonar and an AN/SQQ-61 high-frequency sonar in one automated system. AN/SQQ-90 also includes an AN/SQR-20 towed array sonar. *(General Dynamics Bath Iron Works)*

able targets. This band is also provides target-illumination capabilities for the ship's ESSM – and other surface-to-air missiles that may be installed – dispensing with the need for separate illuminators found on existing cruisers and destroyers. Installation of the radar is designed to preserve the *Zumwalt* class's low RCS, with three flat transmitting and receiving phased-array antenna faces that provide 360° coverage.

The SPY-3's software also has been modified in an attempt to regain some of the volume-search capabilities lost with the deletion of the S band array. The modification allows the radar to alternate between volume and horizon searches at the operator's preference, although optimising the SPY-3 for one type of search can degrade the radar's capabilities in the other mode. The software alteration underwent testing at the Wallops Island Integration and Test Center in Virginia, and will be further evaluated on board the Navy's Self-Defense Test Ship. However, current 2016 configuration does not provide for area air or ballistic missile defence operations.

Sonar: The AN/SQQ-90 integrated undersea warfare suite onboard the *Zumwalt* class includes the US Navy's first dual-frequency hull-mounted sonar system. It consists of three major components: a hull-mounted AN/SQQ-60 mid-frequency sonar, a hull-mounted AN/SQQ-61 high-frequency sonar array, and an AN/SQR-20 multi-function towed-array sonar. Together with embarked MH-60R Seahawk helicopters, the periscope-detection capability resident in the SPY-3 radar, and the inherently quiet operation of the class's electric propulsion, the SQQ-90 provides the new destroyers with significant anti-submarine warfare potential.

The difficult underwater environment in the littorals, where shallow-water sound reverberations can play havoc with detection and detection ranges, was the major driver for the SQQ-61 hull-mounted array. The sonar array also will provide the DDG-1000s with an 'in-stride' mine-avoidance capability. And for its part, the AN/SQR-20 multi-function towed array sonar and handling system, which is capable of both active and passive operation, promises better coverage and detection ranges, as well as less 'down time' than previous towed arrays.

Commercial-off-the-shelf components, automation and more advanced information processing also allow the SQQ-90 to be operated and maintained with up to one-third fewer people than in service fleet sonar systems. SQQ-90 also benefits from the assembly and integration of its sonar electronics into an electronic modular enclosure (EME). The pre-tested EME also provides cooling, shock, vibration and electromagnetic interference protection.

This March 2016 view of *Zumwalt* shows how the arrays for the AN/SPY-3 radar and other sensors are mounted flush with the sides of the deckhouse. Other key features visible include the angled doors in the stern for the internal boat bay located under the large flight deck. The raised safety barriers are fitted in the place of the netting used on other surface combatants and have a radar-absorbing coating to reduce signature. They are raised and retracted automatically. Also of note is the position of the two navigation radars on the superstructure; these will be replaced by two Mk 46 30mm guns when the ship completes fitting out at San Diego. *(Christopher P Cavas)*

A mock-up of the layout of the Common Display System workstations in the *Zumwalt* class's Ship Mission Center (SMC). The Common Display System consoles are identical, three-screen workstations that can be reconfigured to perform a wide variety of tasks necessary for the *Zumwalt's* effective operation. *(Raytheon)*

Total Ship Computing Environment: The sonar EME is just one of several of these modular components that are intended to facilitate shipyard installation and reduce size and weight impacts on the *Zumwalt* class hulls. In total, DDG-1000 will have sixteen EMEs, with each shock-resistant enclosure acting as a data centre holding and supporting the key electronics for all major systems on board the ship. Distributed around the ship, the EMEs then feed the Total Ship Computing Environment (TSCE), an open-architecture, Linux-based information network that ties together the combat management, engineering control, bridge, navigation systems and damage-control systems. In essence, the TSCE provides an internal Internet network for the ship.

This arrangement allows the *Zumwalt* class's systems to be controlled from similar consoles located throughout the ship. The Common Display System features identical, three-screen workstations that can be configured rapidly to allow any type of operation required. In the Ship's Mission Center (SMC) – the DDG-1000 equivalent to the Combat Information Centre (CIC) on older warships – the new displays will allow the heavy, dedicated consoles common to CICs in older ships to be foregone. Instead, the *Zumwalt* class's SMC spaces can be rapidly reconfigured as needed. The CDS consoles are complemented by large flat panel displays in the forward portion of the SMC.

The SMC itself is part of a huge, two deck-high command and control space located in the second and third levels of the superstructure aft of the bridge. There are nineteen Common Display System (CDS) consoles in four rows; the first two being allocated to war-fighting roles; the third to command and control functions and the fourth to engineering and systems support. Above and to the rear of the SMC is a large, glassed-in high deck for mission planners and command staff. Additional planning spaces – again equipped with CDS consoles – are also incorporated into the design.[23]

The new hardware of the CDS is efficient and ergonomically advanced, but software is the true heart of the TSCE. The design and programming effort required to bring the system into existence has been one of the most complex in navy history. As of May 2016, there have been seven versions of TSCE software, which all told incorporates more than eight million lines of code. The eighth variant was scheduled for approval in 2016.

Automation and Optimal Crewing: A key focus of TSCE has been to increase the operational efficiency of the *Zumwalt* class ships. In the surface warship world, as in many other areas, reduced US Navy budgets have translated into the need to reduced ship life-cycle costs, the largest component of which is manpower. The strategic use of automation to reduce crew size has been a key factor in the DDG-1000 design. Unlike past, unsuccessful navy efforts at reducing crew sizes, however, the human system engineering required to make these reductions intelligently has been 'baked into' the ship's top-level design considerations from the start. These take into account shipboard work, maintenance, combat, and damage control constraints that place limits on when and where manpower can be reduced.

Initial US Navy targets aimed for a DDG-1000 crew size of ninety-five, but subsequent analysis determined that that number was too aggressive. With time, the proposed crew size grew first to 114 and then to 148. Ultimately, that settled at 147, which does not include an aviation detachment of twenty-eight people that will join the ship for training and deployment. Still, the overall manning figure for DDG-1000 compares favourably with that of an in-service DDG-51 Flight IIA destroyer.

Ship designers relied on automation in numerous areas to keep manpower levels in check on the *Zumwalt* class ships. To reduce the potentially large numbers of people needed for damage control, for example, they turned to the Autonomic Fire Suppression System (AFSS). This system combines sensors, cameras and automated fire-sensing and firefighting capabilities. AFSS also makes extensive use of 'smart valve' technology that automatically isolates damage in fluid systems (including the chilled water system and the fire mains carrying water used to fight shipboard fires) and re-routes flows using data sent by the valves themselves. Virtual rather than manual monitoring of potential flooding and fire threats in spaces such as the ships' magazines is another method used to contain manpower.[24]

Many of the tasks associated with the manning and monitoring of individual systems – and even controlling the ship itself while underway – also have been automated. The previously-mentioned SQQ-90 sonar suite requires only three people to operate and support it, compared with the six needed to man its predecessor, the SQQ-89. The ship's engineering plant is controlled by only two people, one of whom – the engineering officer of the watch – sits in the SMC from where he or she can keep tabs on conditions in three main machinery and four auxiliary machinery rooms. Underway, DDG-1000's bridge likewise is manned by only three officers.

The navy expects that network automation will reduce maintenance demands on the *Zumwalt* crews. Moreover, as the navy has done with its Littoral Combat Ships, some of the maintenance burden will be transferred from the DDG crews to shore-based teams, so that 'DDG-1000 will require and have several weeks each quarter dedicated to planned and continuous maintenance availabilities', as Captain Kirk explained. 'Consequently, quarterly and above-preventive maintenance and repairs will be done by off-ship maintainers.'[25]

While DDG-1000 brings together numerous efforts to reduce manning and increase the efficiency of smaller crews, not everyone is sanguine that this endeavour will lead to a crew size that is effective in combat. Noting the *Zumwalt*'s plethora of new systems and design features, the US Defense Department's Director of Operational Test and Evaluation cautioned in 2016: 'These systems and equipment have not been subjected to shock on previous ship classes. Moreover, the previously untried automation and small crew for a ship this size, limit the sailors' ability to conduct repairs to enable recovery from shock-induced damage.'[26] The director's statement reflects the perennial trade-off between critical manpower cost saving – with people costs comprising some sixty per cent of the total ownership costs of a ship – and the equally critical need for a truly optimal (as opposed to a minimal) crew size that can fight and save a ship damaged in combat. The US Navy believes that this time it has the number right, although only operational experience will reveal whether or not this is correct.

Two views of one of the two GE Power Conversion advanced electric motors that draw electrical energy from the *Zumwalt* class's integrated power system to turn the ship's fixed propeller shafts. The images show the heavy electrical cables connecting the huge motor to the ship's electrical grid and the direct connection between the motor and the propeller shaft: there is no need for a traditional gearbox. *(Christopher P Cavas)*

Boats and Aviation: The *Zumwalt*'s size provides ample space for the operation and support of off-board systems such as aircraft and boats. The DDG-1000 class ships will be able to carry and support two MH-60 helicopters for many missions. Alternatively, the ship can carry one MH-60 and three MQ-8 series Fire Scout unmanned aerial vehicles, which also take off and land vertically. The size of its flight deck – almost twice that of the DDG-51 class destroyers, and 20ft (6m) higher above the waterline – also increases the safety of flight operations.

The *Zumwalt* class ships also possess an internal boat bay located under the flight deck. This space has room for two 7m or 11m rigid hull inflatable boats (RHIBs), aligned heel-to-toe. The boats are launched and recovered via a combined extendable stern boat ramp and cradle system. When not in use, the bay is protected by a sliding watertight door.

These boat and aviation capabilities – which reflect the original littoral emphasis of the DDG-1000 design – could be particularly useful in supporting special operations teams from the US Marine Corps or other services. Moreover, each *Zumwalt* will have dedicated berthing and planning spaces for these forces.

Power and Electrical System: One of the greatest assets of the *Zumwalt* class ships is the enormous amount of electrical power they will be able to generate – significantly more than any other in-service American naval surface warship. Two Rolls-Royce MT30 generator sets each provide 35.4MW of electrical power, while two RR4500 auxiliary turbine generator sets (each built around a MT5S turbine) each contribute another 3.9MW. Combined, therefore, these main and auxiliary generator sets can deliver 78MW of total ship power. The generators, along with a ship-wide electrical grid and its controls, comprise the class's integrated power system (IPS).

The MT30 gas turbines at the heart of the IPS are known entities. They are a derivative of the

Two views of the internal boat bay on *Zumwalt*, which is located under the flight deck. It can house up to two 11m RHIBs, each on its own lifting cradle. Its inclusion in the design is a reflection of the class's origins as a specialised littoral combatant and the desirability of being able to support Special Forces operations. The ship's design also includes separate accommodation for such personnel. *(Christopher P Cavas.)*

Two images of *Zumwalt* departing harbour for sea trials in very different weather conditions. The ship's integrated power system provides electricity for propulsion, 'hotel' and – potentially – future weapon systems such as railguns and lasers. *(US Navy)*

Trent 800 engines developed to power some Boeing 777s. MT30s already serve the US Navy's *Freedom* (LCS-1) Littoral Combat Ship variants, and will power the British Royal Navy's *Queen Elizabeth* class aircraft carriers and Type 26 frigates and South Korea's FFX Batch II frigates, as well as the Italian Navy's forthcoming LHD type amphibious assault ship.

In the DDG-1000 class, power from the IPS is used to turn two fixed propeller shafts – capped by two fixed-pitch propellers – via two 33.6MW advanced induction motors. These motors, which together can propel the ship to speeds greater than 30 knots, avoid the need for a physical connection between the turbines and the shafts. The IPS also routes power to ship systems and services through the Integrated Fight Through Power (IFTP) system. This distributes DC power throughout the ship, and allows any available generator to meet any demand for current. The system also will reconfigure itself automatically if damage occurs to any distribution element. The engineering control system (ECS) oversees this process – along with auxiliary and some damage control systems – and facilitates centralised system monitoring and management. The ECS, in turn, is below the ship domain controller and command and control system in the computer architecture hierarchy – the latter two systems set the 'guidelines' that the ECS follows as it routes power around the ship.

Overall, the ability to flexibly allocate power between propulsion and systems reduces the number of generators required, increases design flexibility, reduces costs and increases the ships' ability to fight if damaged. These are critical features of the IPS, but it is the sheer amount of electrical power that the *Zumwalt* class ships can produce that has captured much of the US Navy's attention. While steaming at 20 knots, for example, the *Zumwalt* still retains 58MW of reserve power. This is energy that can be used, for example, to power advanced naval weapons now appearing on the horizon.

POTENTIAL GROWTH

In recent years, Department of Defense officials have watched with alarm as Chinese and Russian 'peer competitors' have introduced new weapons and systems that approach – and sometimes equal or surpass – the capabilities of corresponding American systems. The department is looking to 'identify and invest in innovative ways to sustain and advance America's military dominance for the 21st century' to counter the erosion of the longstanding US technological advantages.[27]

The electromagnetic railgun (EMRG) and directed-energy weapons such as solid-state lasers are among the futuristic weapon systems that may contribute to this effort. In-service shipboard defensive and offensive missile and gun systems are available only in limited numbers because of cost and magazine-space constraints. Moreover, many of these weapons often cost many times more than the targets they are designed to kill, setting up an economically-unsustainable contest between the 'expensive and fewer' and the 'many and cheaper.' According to some analysts, even the cost of the AGS LRLAP round has increased significantly as a result of decreased purchases attendant with the truncation from thirty-two to three ships. Both the EMRG and lasers, on the other hand, cost far less than a guided missile: according to one estimate less than US$1/shot for a laser and around US$25,000 for a hypervelocity projectile launched by railgun. Both could theoretically allow a ship to sustain an engagement far longer than current weapons.

The US Navy's EMRG is currently in the second phase of a development programme that began in 2005. Researchers have already achieved 32 megajoule muzzle energy, which – if applied to a shipboard weapon – would be capable of hurling a projectile 100nm (186km) or more. Likewise the AN/SEQ-3 Laser Weapon System has been tested and deployed operationally onboard the interim Afloat Forward Staging Base *Ponce* (LPD/AFSB-15I), in the Persian Gulf. That laser has a 30-kilowatt output, while future naval laser weapons look to generate 100 or more kilowatts.

The US Navy is aiming to use these or similar weapons on future surface warships, including a new, large warship class that could enter service sometime between 2025 and 2030. To field these weapons, these new ships also will be required to generate prodigious amounts of shipboard energy, something that the *Zumwalt* can already do today. Hence, the new DDG-1000 class is a natural choice to test these new weapons as they become available. As Rear Admiral Thomas Rowden, former director of surface warfare (N96), put it, 'We are going to cut our teeth on DDG-1000 and on the IPS we have on that ship … I think a lot of the technologies that we are putting on 1000 to 1002 are going

An image of *Zumwalt* at speed during sea trials in April 2016, illustrating how the ram-like bow cuts through the sea. The ship uses an integrated electrical propulsion system to power two fixed shafts and is capable of speeds in excess of 30 knots. *(General Dynamics/Bath Iron Works)*

to serve us very well as we drive toward these [new] ships.'[28]

But since *Zumwalt* herself is close to entering service, the last two ships in the class, *Michael Monsoor* (DDG-1001) and *Lyndon B. Johnson* (DDG-1002) are the most likely candidates for new-weapon testing. The navy announced in 2014, that the first prototype EMRG would be temporarily placed on board an Expeditionary Fast Transport (formerly designated as Joint High Speed Vessel) for fiscal year 2016 at-sea tests. Longer term, however, the service is hoping to mount an operational railgun on board DDG-1002 – replacing the number two AGS – before the warship makes its first deployment.

A picture of *Zumwalt* taken during initial sea trials in December 2015. She was finally delivered to the US Navy on 20 May 2016 and will be commissioned in October of the same year. *(US Navy)*

Rear Admiral Peter Fanta, in 2016 the director of surface warfare, warned that much engineering and testing remains to be done to determine if this is feasible, but that is the US Navy's goal.[29] The service has no similar plans yet for high-power, solid-state laser testing. But if and when such a weapon becomes available, the *Zumwalt* class ships will again be an obvious choice for their operational testing and deployment.

UNDERWAY, FINALLY!

Zumwalt's fabrication began in February 2009. The ship was launched on 28 October 2013 and BIW was initially scheduled to deliver her to the US Navy in the summer of 2015. However, as is frequently the case with first-in-class vessel construction (and even more often the case with complex warships), that delivery schedule slipped. Instead, the Navy took ownership of *Zumwalt* on 20 May 2016, after a final set of acceptance trials the previous month. It plans to commission the ship into the fleet at a ceremony in Baltimore in October 2016. When delivered, DDG-1000 had not yet been fitted with all elements of her combat system and sensors. Combat system activation will

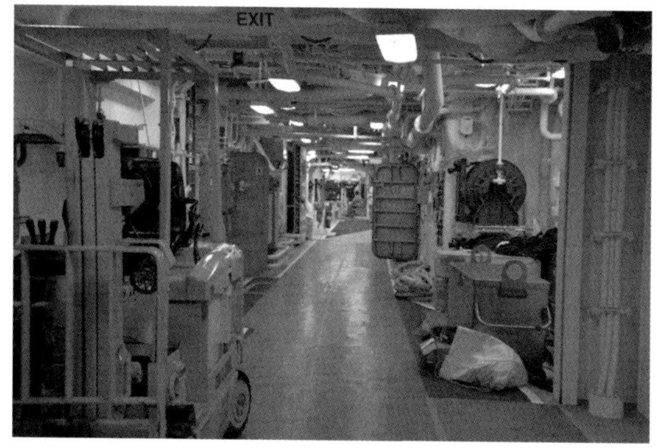

The *Zumwalt* class's substantial size and internal volume will facilitate future upgrades, including the introduction of new weapons systems. This image shows the 'Broadway', an extra-wide passageway running along the ship's main deck on the starboard side from the hangar to the forward magazines. As can be seen, it is broad enough for a fork-lift truck to be driven along. *(Christopher P Cavas)*

Notes

1. Captain George V Galdorisi, US Navy (ret.) and Scott C Truver, 'The *Zumwalt*-Class Destroyer: A Technology "Bridge" Shaping the Navy after Next,' *Naval War College Review – Summer 2010*, Vol 63, No 3 (Newport RI: Naval War College Press, 2010), pp.63–72.

2. See further Norman Friedman, *U.S. Destroyers: An Illustrated Design History* (Annapolis, MD: US Naval Institute Press, 2004), pp.431–5; and Robert O Work, *Know When to Hold 'Em, Know When to Fold 'Em: A New Transformation Plan for the Navy's Surface Battle Line* (Washington, DC, Center for Strategic and Budgetary Assessments, 2007), p.12. A copy of the latter document can be found by searching the CSBA website at www.csbaonline.org. Work has subsequently served as Under Secretary of the US Navy (2009–13) and as of 2016 was the United States Deputy Secretary of Defense.

3. Work, *ibid.* It was decided that around seventy percent of the surface fleet would be full-capability multi-mission combatants, initially the existing *Ticonderoga* (CG-47) and *Arleigh Burke* (DDG-51) classes. The remaining ships would be limited-capability, mission-focused combatants, what became DD-21. After completing the DD-21 programme, the aim was to start replacing the CG-47 class with the second ship of the SC-21 family, a multi-mission CG-21.

4. Rear Admiral Daniel J Murphy, 'Like Thunder and Lightning', *Proceedings* – vol. 123/6/1,132, June 1997 (Annapolis MD: US Naval Institute Press, 1997).

5. Captain Charles S Hamilton, USN, Arsenal Ship Program Manager, *DARPA—Arsenal Ship Lessons Learned* (Arlington VA: Defense Advanced Research Procurement Agency, 1997).

6. Work, *op.cit.*, p.15.

7. Murphy, *op.cit.* Some sources suggest that as many as 256 VLS cells were initially planned.

8. Ronald O'Rourke, *Navy DD(X) and LCS Ship Acquisition Programs: Oversight Issues and Options for Congress RL32109* (Washington, DC: Congressional Research Service, 2005), pp.10–11.

9. Murphy, *op.cit.*

10. O'Rourke, *op.cit.*, p.14.

11. Admiral Gary Roughead, Remarks to the Surface Navy Association Symposium, 14 January 2009.

12. The Nunn-McCurdy provision requires a number of actions to be taken to avoid termination if a major defence programme exceeds certain cost-growth parameters.

13. Ronald O'Rourke, *Navy DDG-51 and DDG-1000 Destroyer Programs: Background and Issues for Congress RL32109* (Washington, DC: Congressional Research Service, 12 June 2015), pp.33–5.

14. Captain James Kirk, US Navy, 'Ahoy from the *Zumwalt*', *Proceedings* – vol. 142/3/1357, March 2016 (Annapolis MD: US Naval Institute Press, 2016), p.21.

15. Christopher P. Cavas, 'Is New U.S. Destroyer Unstable? Experts Doubt Radical Hull; Navy Says All is Well', *Defense News* – 2 April 2007 (Springfield VA: Sightline Media Group, 2007). A copy can be found by searching www.defensenews.com.

16. Grace Jean, 'USN Presses on with *Zumwalt* class Destroyers', *ISS Jane's Navy International* – 19 January 2016 (Coulsdon: IHS Janes, 2016).

17. Michael R LeGault, 'DDG-1000 *Zumwalt*: Stealth Warship', *Composites World* – 18 January 2010 (Cincinnati OH: Gardner Business Media, Inc., 2010).

be conducted at the ship's homeport of San Diego, California, a decision the navy made to free up space and capacity at BIW, which also is dealing with a thriving DDG-51 construction programme.

The same delay, which a US Navy spokeswoman attributed to 'the challenges encountered with completing installation, integration and testing of the highly unique, leading edge technology designed into this first-of-class warship', has also pushed back DDG-1001's planned launch and delivery schedule. However, that ship was approximately eighty-four percent complete in January 2016, according to the service. She was christened on 18 June 2016.

Ultimately, the last two ships in the class will follow in *Zumwalt*'s wake and will be homeported in San Diego. The basing of all three ships on the West Coast seems to be a direct consequence of the United States' 'Pivot to the Pacific' – a rebalancing of American forces and greater attention to the Indo-Pacific basin, announced by President Barack Obama in 2011 – in response to perceived challenges from China and the rapid growth of Chinese military capabilities.

Ironically, the Pacific theatre may not offer the DDG-1000-class ships a significant opportunity to test their littoral-combat and fire-support capabilities. But the Pacific basin is a theatre where the future of naval warfare – one dominated by new technologies and capabilities – may take shape. Given the programme's significant overall cost and its truncation to just three ships, the *Zumwalt* class's ultimate success may lie less in the capabilities planned for them in the 1990s, and more in their ability to 'midwife' the technologies the Navy will need throughout much of the twenty-first century.[30]

The *Zumwalt* class's large reserves of electrical power make them eminently suitable for testing new generations of US Navy weapons. One of these is an electromagnetic railgun (EMRG), which has already achieved 32 mega-joule muzzle energy – sufficient for a hundred mile range. These images show the latest 32-MJ version of the weapon and a test of an earlier, lower-powered version. The flame is from the ignition of particle debris. Current plans envisage one of these weapons being installed aboard *Lyndon B. Johnson* (DDG-1002), the third and final member of the *Zumwalt* class. It is *Zumwalt*'s ability to 'midwife' such future technology that provides its real value to the fleet. *(US Navy)*

18. Sam LaGrone, 'New External DDG-1000 Mast Reduces Ship's Stealth From Original Design', *USNI News* – 3 March 2016 (Annapolis MD: US Naval Institute, 2016). A copy can be found by searching www.news.usni.org/. The changes are not immediately apparent from early views of the ship on trials but much communications equipment will be fitted only after the ship has been formally delivered.

19. Northrop Grumman decided to spin off its shipbuilding entities – principally the Ingalls and Newport News shipyards – into a new listed group known as Huntington Ingalls Industries in 2011.

20. Sam LaGrone, 'Navy's Steel Deckhouse Decision for Final *Zumwalt* is a Blow to HII', *USNI News* – 5 August 2013 (Annapolis MD: US Naval Institute, 2013).

21. The US Navy has subsequently put into service a longer-range Mk 45 Mod 4 62-calibre 127mm gun.

22. Sam LaGrone, 'Navy Swaps Out Anti-Swarm Boat Guns on DDG-1000s', *USNI News* – 14 August 2014 (Annapolis MD: US Naval Institute, 2014). The 57mm Mk 110 guns are incorporated into both Littoral Combat Ship variants as well as the US Coast Guard's National Security Cutters and the decision was greeted with some surprise. Some commentators suggest that cost and weight-saving considerations – rather than overall effectiveness – were key to the decision.

23. Christopher P Cavas, 'At Sea Aboard the *Zumwalt*,' *Defense News* – 31 March 2016 (Springfield VA: Sightline Media Group, 2016).

24. Captain George V Galdorisi, US Navy (ret.), and Scott C Truver, 'From Minimal to Optimal', *Proceedings* – vol. 137/7/1,301, July 2011 (Annapolis MD: US Naval Institute Press, 2011), p.27.

25. Kirk, *op.cit.*, p.22.

26. Director of Operational Test & Evaluation, *FY 2015 Annual Report* (Arlington VA: US Department of Defense, 2016), p.190.

27. Deputy Secretary of Defense Robert O Work, 'The Third U.S. Offset Strategy and its Implications for Partners and Allies', speech, Washington, DC, January 28, 2015.

28. 'In Pursuit of the U.S. Navy's Next Surface Combatant', *USNI News* – 14 January 2014 (Annapolis MD: US Naval Institute, 2014). As Commander Naval Surface Forces, Vice Admiral Rowden has championed increased, distributed lethality throughout the fleet.

29. Christopher P Cavas, 'Navy's Rail Gun Still Headed to Sea, but on Which Ship?', *Defense News* – 10 January 2016 (Springfield VA: Sightline Media Group, 2016).

30. Cost data on the *Zumwalt* programme continues to evolve but currently amounts to significantly over US$20bn. For further detail see Jacob Cohn and Jesse Sloman, *FY2016 Weapon Systems Factbook* (Washington, DC: Center for Strategic and Budgetary Assessments, 2016), p.65; Sam LaGrone, 'Navy Requires $450 Million More to Complete *Zumwalt*-Class Due to Shipyard Performance', *USNI News – 6 April 2016* (Annapolis MD: US Naval Institute, 2016). As of FY2015, US$20.8bn had been appropriated to design and construct the *Zumwalt* class and another US$1.2bn was required to complete the programme. Since then costs have increased by around US$0.5bn and the US Navy has indicated a further US$0.4bn is required for post-delivery expenses outfitting. Of this total of c. US$22.9bn, US$9.7bn relates to research and development, US$3.8bn for DDG-1000, US$2.8bn for DDG-1001, US$2.4bn for DDG-1002, US$3.8bn for one-off engineering work and US$0.4bn for the post-delivery work already mentioned.

4.1 TECHNOLOGICAL REVIEW

Author:
David Hobbs

WORLD NAVAL AVIATION

AN OVERVIEW OF RECENT DEVELOPMENTS

2016 is seeing a measure of stability with respect to naval aviation programmes across many of the world's navies. New ships are entering service as planned, new generations of aircraft are reaching initial operational capability (IOC), and the interest in deploying affordable aviation support ships has been reflected in growing numbers of amphibious 'flat-tops'. The US Navy has begun to investigate replacement platforms to maintain the capabilities of Super Hornets and Growler jets, as well as of Sea Hawk helicopters, after 2030 and is paying particular attention to unmanned combat air vehicles. The revival of the United Kingdom's carrier strike capability has moved nearer with the projected handover of the first CVF aircraft carrier, *Queen Elizabeth*, in 2017, significant progress made on her sister-ship *Prince of Wales*, the formation of F-35B squadrons in the United States and continuing work to upgrade Merlin helicopters serving in three distinct roles. An overview of the impact of the 2015 Strategic Security and Defence Review (SDSR) and the current status of British maritime aviation concludes this annual review.

An F/A-18F two-seat Boeing Super Hornet strike fighter is launched from the flight deck of *John C. Stennis* (CVN-74) whilst the carrier was operating in the South China Sea in May 2016. Delays to the arrival of the carrier F-35C variant of the Lightning II Joint Strike Fighter mean that there is a risk the current strike fighter fleet will 'wear out' before replacement; production of Super Hornets is therefore being extended as one of a number of mitigants. *(US Navy)*

US NAVY: AIRCRAFT CARRIERS AND THEIR AIR WINGS

The US Navy's carrier fleet will return to its mandated level of eleven vessels by the end of 2016 with continued funding for the mid-life refuelling and complex overhaul (RCOH) of older ships and the new *Gerald R. Ford* (CVN-78) expected to join the fleet in the autumn. Under the 2014 Optimized Fleet Response Plan (O-FRP), two carriers are normally deployed at any one time, with others capable of deployment at short notice.[1] Studies into less expensive carrier designs continue but it is hard to escape the conclusion that nuclear carriers, with their versatile air wings, remain the most reliable and cost-effective means of projecting both 'hard' and 'soft' power across the globe. A larger number of smaller, individually less capable ships – an option that has its supporters in Washington – could well prove to be a more expensive way of delivering the same effect.

This conclusion was also the essential outcome of a research paper published by the Hudson Institute in October 2015, which argued that big carriers

The US Navy *Nimitz* class aircraft carrier *Dwight D. Eisenhower* (CVN-69) pictured undertaking pre-deployment training in the Atlantic on 29 March 2016. The American carrier fleet has been operating under a degree of strain for some time but delivery of the new *Gerald R. Ford* (CVN-78) before the end of 2016 will return the force to its approved eleven-ship strength. *(US Navy)*

would not only be an essential factor in any conflict with a peer or near-peer adversary but that they would also be the best option available.[2] To be fair, the Institute criticised the limited strike radius of contemporary air wings that followed a focus on low-intensity conflict after the end of the Cold War as a shortcoming that must be addressed as a wider range of anti-access/area denial (A2/AD) weapons become available to potential enemies.[3] It also observed that the US Navy's logistic supply vessels would be inadequate to support a sustained campaign in the Pacific and that the fleet relies too heavily on active electronic emissions for command and control of its operations. However, the Institute emphasised the inherent ability of carrier task forces to loiter near the edges of an enemy's A2/AD volume, strike at critical targets and then re-position at high speed to different operating areas. These 'power pulse' or 'hit and move' operations would be difficult for any A2/AD system to counter. The Institute concluded that, far from diverting money to potential alternatives, the navy should be allowed to get the best out of the big-deck carriers that it already has, especially 'buying back' the ability of air wings to strike at targets beyond 1,000nm (1,850km) from the carrier.

Logical measures to reduce carrier construction costs are, however, being considered. For example, the Aircraft Carrier Industrial Base Coalition – a lobby group representing businesses that supply parts, equipment and services for carrier construction and maintenance – recently briefed the US Naval Institute (USNI) that it is asking for US$293m to be pulled forward into the FY2017 budget to support the advanced procurement of equipment for CVN-80, the future *Enterprise*, and the as yet un-named CVN-81. Purchasing equipment in blocks is a proven way of lowering cost through economies of scale and the same arguments would apply, equally, to the advance procurement of materials for the RCOH of *John C. Stennis* (CVN-74) while work is underway on *George Washington* (CVN-73).

Meanwhile, the anticipated 'fighter gap' that will arise in the mid-2020s, as a result of the delayed F-35C Lightning II failing to replace 'legacy' F/A-18A/B/C/D Hornets in a timely fashion as they reach the end of their 6,000-hour fatigue lives is a major US Navy concern. One solution is an ongoing service life extension programme (SLEP) for the 'legacy' Hornets to extend their fatigue lives to 8,000 or perhaps even 10,000 hours. After survey, suitable aircraft are taken into aviation depots but no two require the same re-work and a 'logjam' has prevented work being carried out on other important tasks. Throughput is improving, however, and greater capacity is expected in 2017. The expanded use of F/A-18E/F Super Hornets to replace 'legacy' Hornets has filled the gap in the short term but – added to the number of operational missions flown against targets in Iraq, Afghanistan and Syria over the past decade – this has meant that Super Hornet hours are being used up considerably faster than originally predicted. Indeed, some early-production aircraft are already reaching the end of their fatigue lives, making a Super Hornet SLEP inevitable. Another step is to continue F/A-18E/F construction until the F-35C can enter service in the necessary numbers.

Indeed, the US Navy's budget request for 2017 – detailed in Table 4.1.1 – seeks to keep the Super Hornet production line open. Funding is requested for two F/A-18E/Fs in 2017, with a further fourteen in 2018. Other requests in the US$14.1bn proposal include four F-35C Lightning II Joint Strike Fighters, six E-2D Advanced Hawkeye early warning aircraft, eleven P-8A Poseidon maritime patrol aircraft, and one MQ-8C Fire Scout, two MQ-4C Triton and eight RQ-21A Blackjack unmanned aerial vehicles. When combined with additional aircraft added by Congress to the FY2016 budget request, there is a clear trend towards higher purchases of naval aircraft than previously envisaged, particularly in the fast jet category.

A further measure is the proposed deactivation of Carrier Air Wing (CVW) 14 which has not embarked in a carrier since 2012 and has recently been in 'cadre' status. The 2016 budget had contained proposals to restore it to full operational status but Vice Admiral Mike Shoemaker, Commander Naval Air Forces, described the de-

Table 4.1.1: US NAVY PLANNED AIRCRAFT PROCUREMENT: FY2016–FY2021

TYPE	MISSION	FY2016[1]	FY2017[2]	FY2018	FY2019	FY2020	FY2021	FYDP 2017–21[3]
Fixed Wing (Carrier Based)								
F-35B Lightning II JSF	Strike Fighter (STOVL)	15	16 (14)	20 (20)	20 (20)	20 (20)	21	**97 (89)**
F-35C Lightning II JSF	Strike Fighter (CV)	6	4 (4)	6 (8)	12 (10)	18 (12)	24	**64 (40)**
FA-18E/F Super Hornet	Strike Fighter (CV)	5	2 (0)	14 (0)	0 (0)	0 (0)	0	**16 (5)**
EA-18G Growler	Electronic Warfare	10[4]	0 (0)	0 (0)	0 (0)	0 (0)	0	**0 (10)**
E-2D Advanced Hawkeye	Surveillance/Control	5	6 (6)	5 (5)	3 (4)	4 (4)	5	**23 (24)**
Fixed Wing (Land Based)								
P-8A Poseidon	Maritime Patrol	17	11 (12)	6 (12)	13 (7)	0 (0)	0	**30 (48)**
C-40A Clipper	Transport	1	0 (0)	0 (0)	0 (0)	0 (0)	0	**0 (1)**
KC-130J Hercules	Tanker	2	2 (2)	2 (2)	2 (2)	2 (2)	2	**10 (10)**
Rotary Wing								
AH-1Z/UH-1Y Viper/Venom	Attack/Utility	29	24 (27)	27 (27)	27 (27)	0 (0)	0	**78 (110)**
VH-92A	Presidential Transport	0	0 (0)	0 (0)	6 (6)	6 (6)	5	**17 (12)**
CH-53K Super Stallion	Heavy-Lift	0	2 (2)	4 (4)	7 (7)	13 (13)	14	**40 (26)**
MV-22B/V-22 Osprey	Transport	19	16 (18)	6 (8)	6 (8)	6 (8)	14	**48 (61)**
MH-60R Seahawk	Sea Control	29	0 (0)	0 (0)	0 (0)	0 (0)	0	**0 (29)**
Unmanned Aerial Vehicles								
MQ-8C Fire Scout	Reconnaissance	5	1 (2)	2 (2)	2 (2)	2 (2)	2	**9 (13)**
MQ-4 Triton	Maritime Patrol	4	2 (3)	3 (4)	3 (4)	5 (4)	6	**19 (19)**
RQ-21A Blackjack	Tactical Reconnaissance	6	8 (4)	4 (4)	5 (5)	5 (5)	3	**25 (24)**
Totals:		**153**	**94 (94)**	**99 (96)**	**106 (102)**	**81 (76)**	**96**	**476 (521)**

Notes:

1 FY2016 base numbers relate to the authorised procurement programme. This varied quite significantly from the initial Presidential budget request for just 124 aircraft, with the difference largely comprising additional orders for F-35 and, particularly, FA-18E/F and EA-18G fast jets.

2 Numbers for 2017 to 2021 relate to base FY2017 budget plans including two F/A-18 and four RQ-21A aircraft requested from overseas contingency operations funds; numbers in brackets reflect purchases for that year previously envisaged in the FY 2016 budget.

3 Future Years Defence Programme; numbers in brackets reflect FYDP for 2016–20 as adjusted for the authorised 2016 programme.

4 Includes three aircraft funded by Boeing as part of a 2014 settlement of a long-running dispute related to the cancellation of the A-12 Avenger programme in the early 1990s.

activation of the wing as being 'the most efficient use of operational resources to meet global requirements'. It would allow, he said, 'predictable rotations between squadrons and reduce excessive time between deployments when carriers are undergoing lengthy periods of maintenance'. CVW-14's deactivation would require three squadrons to be stood down, the formation of one squadron to be cancelled and an electronic attack VAQ Growler squadron to be transferred to a land-based expeditionary role. One of the units that would be disbanded is VFA-15, the 'Valions', equipped with F/A-18C Hornets. This would leave the US Navy with just thirty-four VFA strike fighter squadrons, two short of the number required to equip the nine remaining operational CVWs. Traditionally such shortfalls have been made up by the embarkation of US Marine Corps (USMC) VMFA units and this will continue to be the case. Standing down the 'legacy' Hornets of VFA-15 should help to reduce the pressure from the Hornet SLEP programme on aircraft depots and, in the longer term, reduce the size of the 'fighter gap'.

Meanwhile, the Unmanned Carrier-Launched Airborne Surveillance and Strike (UCLASS) project has been replaced by a programme initially referred to as the Carrier Based Aerial Refuelling System (CBARS). Progress with UCLASS had stalled after disagreements between US lawmakers and the US Navy over the exact definition of the type's mission. CBARS initially retained the RAQ-25 designation from UCLASS but, from April 2016, the US Navy has referred to it as the MQ-XX (reflecting unmanned, multi-mission potential) and has used the name 'Stingray' in briefing documents. However, the Director of Air Warfare (OPNAV N98), Rear Admiral Mike Manazir, briefed the Senate Armed Services Committee in April 2016 that MQ-XX would only include tanking and intelligence, surveillance and reconnaissance (ISR), as primary missions. He also said that the USN 'will only consider non-developmental ISR systems' at first, because these can be achieved rapidly. He explained how the X-47B had proved that unmanned aircraft could take off from and land on a carrier and refuel in the air but that MQ-XX had now to demonstrate that it could 'do two basic meat-and-potato missions' at first. That 'will also provide a platform for us to go forward and do additional, more advanced, capabilities, which could include long-range strike eventually'.

An F/A-18E Super Hornet assigned to the 'Warhawks' of Strike Fighter Squadron (VFA) 97 prepares to refuel an EA-18G Growler assigned to the 'Wizards' Electronic Attack Squadron (VAQ) 133 in February 2016. Between five and six Super Hornets in every embarked carrier air wing are currently used for tanking missions, partly reflecting the relatively short range of the aircraft. The planned MQ-XX unmanned tanker aircraft, which is arising from the ashes of the UCLASS programme, will allow these aircraft to be released for combat roles. *(US Navy)*

A US Navy C-2A Greyhound carrier-on-board delivery aircraft landing on the carrier *Harry S. Truman* (CVN-75) in November 2015. First introduced in 1964, the Greyhound is closely related to the Hawkeye airborne early warning aircraft. It is to be replaced by a variant of the V-22 Osprey, which will be able to land on a much wider range of ships. *(US Navy)*

The Chief of Naval Operations (CNO), Admiral John Richardson, has described how MQ-XX will 'help to ease the burden on the F/A-18E/F Super Hornet fleet by taking over the tanking mission'. Five or six Super Hornets in every air wing are currently used for tanking missions and their replacement would free them for their primary task and also help to reduce the 'fighter gap'. In 2016 the Navy presented the initial MQ-XX requirements to

Although the British Royal Navy has temporarily 'gapped' aircraft carrier operations pending the arrival of the new *Queen Elizabeth*, the British Type 45 destroyer *Defender* has spent considerable time in the company of other nation's carriers over the past twelve months. The first image shows her operating with France's *Charles de Gaulle* in December 2015 during the latter's deployment to the Middle East for strikes against Islamic State forces. *Defender* subsequently participated in the Indian fleet review at Visakhapatnam in February 2016, where she is seen in company with the veteran carrier *Viraat*, the former British Royal Navy *Hermes. Viraat* and her Sea Harriers were withdrawn from service soon after the review had concluded. *Defender* also operated with a third carrier, the US Navy's *Harry S. Truman* (CVN-75) during the course of her deployment. *(Crown Copyright 2015, 2016)*

the Pentagon's Joint Requirements Oversight Council and requests to industry for proposals are expected to follow in 2017. Four companies are expected to bid for the work: General Atomics, Boeing, Lockheed Martin and Northrop Grumman.

Given the direction MQ-XX is currently taking and the relatively modest number of F-35C variants being procured, it has become clear that the US Navy will still struggle to provide the capabilities currently provided by Super Hornets and Growlers into the 2030s. Attention has, therefore, begun to focus on potential replacements, referred to initially as the F/A-XX programme. However, in January 2016, the US Navy announced that – rather than a single platform – several systems designed to fill the strike and air warfare missions in carrier air wings would be required. The project has therefore been re-named as Next Generation Air Dominance (NGAD). Rear Admiral Manazir briefed the USNI that the navy initially considered partnering with the US Air Force (USAF) in a joint project to replace the F-22 Raptor and F/A-18E/F but has now decided not to seek a common platform. It will, though, 'utilise common analytical tools and methods and will share information' across a number of studies; an approach designed 'to ensure interoperability and leverage opportunities for common technology investments'.

The US Navy's approach to NGAD was approved by the Joint Chiefs of Staff in 2016. Work is now in hand to define the requirement, which may take the form of evolutionary or incremental platform development, derivations from legacy platforms or all-new designs. The emphasis will be on replacing the capabilities of Super Hornets and Growlers rather than being wedded to a one-for-one replacement of the airframes. Individual system elements might be manned or unmanned and, above all, must take on board hard-won lessons from the overly-optimistic F-35 programme. The previous CNO, Admiral Jonathan Greenert, stressed that what is now known as NGAD should deploy a wide spectrum of weapons and penetrate hostile airspace by suppressing enemy air defences rather than attempting to hide from them. Interestingly, the US Navy appears to place more emphasis than the USAF on the development of unmanned systems. Secretary of the Navy Ray Mabus has previously stated publicly that the F-35C could and should be the last manned US Navy strike fighter, with 'unmanned systems becoming the new normal in ever-increasing areas'.

In April 2016 the Naval Air Systems Command awarded Bell-Boeing a US$151m contract to begin work on the navy variant of the MV-22 Osprey which is to replace the venerable C-2A Greyhound as a carrier-on-board delivery (COD) aircraft. It is to be designated the CMV-22B, a fusion of the cargo-carrying USAF CV-22 and the multi-mission USMC MV-22. Production is to begin in 2018, with the first delivery in 2020. Ray Mabus told the Senate Armed Services Committee in March 2016 that CMV-22B production will commence as USMC procurement of the MV-22 draws down, enabling the navy to 'fold its procurement into an existing multi-year contract'. Like its predecessor, the CMV-22B is to be based at naval air facilities ashore but has greater flexibility since it can land on both aircraft carriers and big-deck amphibious ships. It can also land on Military Sealift Command supply ships, destroyers and a range of amphibious ships, as well as allied carriers and amphibious vessels.

OTHER AIRCRAFT CARRIER OPERATORS

The French Navy's nuclear-powered aircraft carrier *Charles de Gaulle* demonstrated its ability to integrate with the US Navy in late 2015 when she became flagship of Task Force 50 in the Arabian Gulf, the first time that a French warship has commanded a US Navy task force. She continued in this role until relieved by *Harry S. Truman* (CVN-75) in February 2016. During the deployment she flew 370 operational missions as part of Operation 'Inherent Resolve', including eighty strikes against militants of the so-called 'Islamic State' and twenty-three reconnaissance sorties over Iraq and Syria. After passing back through the Suez Canal she also took part in exercises with Egyptian armed forces. The deployment marked the last operational embarkation of the Super-Étendard strike fighter, which made its final launches from the carrier on 16 March 2016. This leaves French fast jet naval aviation focused on the naval 'M' variant of the Dassault Rafale multi-role fighter.

Chinese media has published details of the nation's second aircraft carrier following an official announcement that it was under construction by defence spokesperson Yang Yujun on 31 December 2015. The China News Service's English-language ECNS reported that the new carrier is to be larger than the existing *Liaoning* but, like her, will operate J-15 strike fighters. She will benefit from experience with the earlier ship, including greater magazine capacity and fuel stowage to decrease the amount of underway replenishment required during prolonged deployments, thereby improving combat effectiveness. The People's Liberation Army Navy Air Force (PLANAF) has confirmed that, in addition to the J-15, the new ship is to operate airborne early warning aircraft, anti-submarine aircraft and helicopters capable of humanitarian relief operations.

The Indian Navy withdrew *Viraat*, formerly the British *Hermes*, from service in 2016, along with their handful of remaining Sea Harrier FRS.51 strike fighters. There are plans to preserve the carrier as a floating hotel and conference centre at Visakhapatnam. Her ship's company moved to stand-by India's first indigenously-designed aircraft carrier, the 40,000-ton *Vikrant* (II), which is due for completion in 2018.[4]

Until then the Indian Navy will have only one operational aircraft carrier, *Vikramaditya*. However, it plans to build a further, possibly nuclear-powered, carrier after *Vikrant*. Foreign carrier designs are being studied before requesting proposals and an agreement has been signed with the United States that would allow the procurement of the electromagnetic aircraft launch system (EMALS) and advanced arrestor gear (AAG) similar to those fitted in *Gerald R. Ford*, together with other aspects of carrier technology. There seems to be little interest in the British *Queen Elizabeth* class design but a French delegation has reportedly visited India with proposals for a DCNS derivative, fitted with steam catapults and Mk 13 arrester gear.

Surprisingly, considering the late and over-budget delivery of *Vikramaditya*, India is also reported to be interested in a new Russian design known as Project 23000E 'Storm'. Models of the design show a Short Take-Off But Arrested Recovery (STOBAR) capability – using a large 'ski-jump' at the bow – as well as two EMALS-like catapults able to launch aircraft from an angled deck. Arrester wires recover aircraft from both types of launch and there are two separate islands, as in the case of *Queen Elizabeth*. Interestingly, quoted displacement and dimensions are almost identical to those of the new *Gerald R. Ford*. 'Storm' is reportedly capable of remaining at sea for 120 days with a ship's company of up to 5,000 and an air wing of up to 100 aircraft. These would include a naval derivative of the T-50 (PAK FA) stealth fighter, Mig-29K strike fighters and airborne early warning aircraft, most likely the Yak-44E. The composition of the air wing may appeal to the Indian Navy. It is already the largest user of the Mig-29K and an 'Indian' version of the PAK FA is believed to be under development for the Indian Air Force. Progress with the naval version of the indigenous Indian Tejas light fighter has been slow and a package that includes both new carrier technology and strike fighters could be attractive.

F-35 LIGHTNING II PROGRESS

In testimony before the US House Armed Services Tactical Air and Land Forces Subcommittee in early 2016, witnesses confirmed that F-35 progress reflects growing maturity, with lowering costs and improving repair, maintenance and availability statistics. 2015 saw the declaration of initial operational capability (IOC) for the first F-35 squadron. The US Marine Corps' VMFA-121, the 'Green Knights', based in Yuma, Arizona achieved this milestone on 31 July 2015, using F-35B Short Take Off and Vertical Landing (STOVL) variants with Block 2B software. The squadron is to deploy forward to Marine Corps Air Station (MCAS) Iwakuni, Japan in January 2017 where it will form part of Marine Air Group (MAG) 12. The USMC programme of record specifies the eventual procurement of 353 F-35Bs and sixty-seven F-35Cs to replace the AV-8B Harrier, F/A-18A/B/C/D Hornet and EA-6B Prowler, with the last active-duty squadron to convert in 2031.[5] The final AV-8B Harrier squadron is to stand down in 2025 and the last F/A-18 squadron in 2030. However, both dates are subject to a review in 2019. Meanwhile, the second front-line F-35B squadron, VMFA-211, the 'Wake Island Avengers', exchanged its AV-8Bs for F-35Bs during 2016 and is expected to reach IOC in 2018. Two F/A-18 squadrons are to convert after that, VMFA-122, the 'Werewolves', from 2018 and VMFA-115, the 'Silver Eagles', from 2020. They will be followed in turn by another AV-8B unit which will become VMFA-311, the 'Tomcats', from 2021.

However, development is still far from complete. Lieutenant General Christopher Bogden USAF, the F-35 Programme Executive Officer, has briefed on a number of problems, among them software issues related to the migration of Block 2B software into improved hardware to create Block 3i. Both were highlighted in a 48-page report on the project written by Michael Gilmore, the Pentagon's Director of Operational Test & Evaluation (DOT&E), published as part of the DOT&E's FY2015 annual report.[6] He stated bluntly that the limited capability

Viraat's withdrawal leaves India with only one operational aircraft carrier, the Russian-built *Vikramaditya*, until the indigenous *Vikrant* (II) is completed. *Vikramaditya* is seen here at the Indian Fleet Review in February 2016. *(Hartmut Ehlers)*

An F-35C Lightning II carrier variant pictured embarked on the carrier *Dwight D. Eisenhower* (CVN-69) in October 2015 during 'at sea' trials. Whilst the F-35's development is far from complete, the programme saw material progress during 2015-16. *(US Navy)*

F-35C Lightning II carrier variants and F-18E/F Super Hornets pictured operating together over NAS Fallon's extensive training areas in Nevada. The two aircraft will together form the major part of the US Navy's strike aircraft force from the mid-2020s onwards. Recently announced plans suggest that the US Navy will use each F-35 more extensively than was initially envisaged, possibly reflecting likely ongoing pressure on overall naval fast jet numbers. *(US Navy)*

of Block 2B standard software and technical problems which impose speed and manoeuvre limitations on VMFA-121's aircraft mean that, at present, 'they would need to avoid threat engagement in an opposed combat scenario' unless they were augmented by other friendly forces. Block 3i software reset problems have led to radar blank-outs in US Air Force F-35As but USMC F-35Bs were not loaded with the update at fault and have not been affected. Work to fix Block 3i has slowed work on Block 3F software, the planned IOC standard for US Navy and British F-35s from 2018. Acceptance of Block 3F will mark the end of the Systems Development and Demonstration (SDD) phase of the project but flight testing has started late and Gilmore does not anticipate that it can be completed by the target date of May 2017.

F-35 software involves four functionally-interlinked but distinct elements, viz. (i) the aircraft; (ii) the Autonomic Logistic Information System (ALIS) which manages maintenance, support, flight planning and technical competence; (iii) the reprogramming laboratory complex which develops the mission data file (MDF) software intended to allow fused sensors to identify threats and targets; and (iv) the simulators used for training and mission rehearsal. The need for all four to work seamlessly together grows in importance as the expanding number of deployed operational squadrons require both live and virtual training and support. Every aspect of this structure is, according to Gilmore, delayed and there will likely be no end to development through the life of the aircraft. The possible damage that might be caused to this structure by a successful cyber-attack can easily be imagined.

In addition to software issues, other problems have arisen that have required resolution. For example, during 2015/16 an issue was identified with airframe heating which was thought to have been addressed with a more efficient fuel pump, since the F-35 uses fuel as a heat sink to cool some of the airframe's many electronic systems. Like all development problems, this one will be cured but it is obviously complicated and will take time. Another problem was the unacceptable risk of neck injury found to affect pilots weighing less than 136lbs (61.6kg) when using the US16E seat to eject at lower airspeeds, which emerged during sled testing by Martin Baker. Fixes are to be developed and funded by Martin Baker. A third-generation pilot's helmet reduced in weight from 5.1lbs to 4.6lbs (2.5kg to 2kg) to reduce neck loads during ejection will also help to resolve the problem.

Meanwhile, the US Navy has increased its projected annual use of F-35C Lightning IIs from 300 to 360 hours per airframe after a study which balanced potential simulator training against live, virtual and constructive (LVC) training. This brings the navy broadly into line with the USAF's planned 350-hour usage (a reduction from the 400 hours previously envisaged). The USAF, USMC and US

Two views of a CH-53K King Stallion heavy-lift helicopter. The advance in capability provided by new amphibious assault ships is reflected in new, more capable aircraft being acquired for the US Marine Corps. The MV-22 Osprey has already transformed personnel transportation, whilst heavy-lift capabilities will be enhanced by the new CH-53K, the latest iteration of CH-53 Sea Stallion series. The new helicopter first flew on 27 October 2015. *(Lockheed Martin)*

Navy have also revised usage rates for the life of the programme, adding 1.6 million flying hours across all three variants between 2016 and 2070. The change is expected to cost the programme an extra $23bn for the continued production of spares and other support requirements but will delay the time-frame under which potential replacements need to be procured.

BIG-DECK AMPHIBIOUS SHIPS AND THEIR AIRCRAFT

The first *America* (LHA-6) class amphibious assault ship will join the operational fleet in 2017 after spending much of 2015 and early 2016 undergoing a work package to strengthen her flight deck and coat it with heat-resistant material to enable her to operate F-35Bs. Catwalks and their contents, including emergency life-rafts, have had to be similarly coated or protected. The second ship, *Tripoli* (LHA-7), is to undergo similar work after her completion, although some deck strengthening has been achieved during build. The projected third ship in the class, the as-yet unnamed LHA-8, will see significant design modification to incorporate a well deck for landing craft and is also to be built from the outset with the stronger flight deck.

In a late 2015 briefing Lieutenant General Jon Davis, the US Marine Corps Deputy Commandant for Aviation, briefed the USNI about his vision for the future. He believes that the US Navy and Marine Corps should explore ways of making USMC aircraft 'a more useful part of the naval battle force, using alternative mixes of aircraft on amphibious ship flight decks' and seeking new missions for them. At present big-deck amphibious ships embark Marine Expeditionary Units (MEUs) with a typical air group comprising six AV-8B Harriers or, in the near future, F-35Bs, twelve MV-22B Ospreys, three or four CH-53E Super Stallions, seven light attack helicopters and several small UAV units, a combination that includes 'a little bit of everything' but which could be criticised for not actually being optimised for any specific mission. Davis believes that there are many scenarios where it would be desirable to have the big-deck amphibious ship loaded with a full squadron of sixteen F-35Bs together with six MV-22Bs fitted with an air-to-air tanking package. That could be, he feels, the way these capable ships might deploy as a matter of course to form part of a naval task force. Some of the helicopters, including the light attack 'skids', could be dispersed to other

amphibious ships or even to destroyers. The big-deck ships' ability to project power with F-35Bs would then augment the carrier strike groups' capability in the opening stages of a high-intensity conflict. When that augmentation was no longer needed, the F-35Bs could be deployed ashore and assault assets – including more MV-22s and the new CH-53K King Stallion – could be flown onto the amphibious carrier utilising their long range and air-to-air refuelling capabilities.

Davis' argument that maximum flexibility should be derived from the amphibious carriers and their 'game-changing' aircraft offers possibilities that were endorsed by the Corps' Commandant, General Robert Neller, in February 2016 when he explained that both he and CNO Admiral Richardson wanted to see amphibious forces participate in a wider range of missions – including long-range strike from the sea, electronic warfare and other condition-setting activities – before the marines go ashore. Lieutenant General Davis has also mentioned the USMC's interest in UCAVs but he noted that MQ-XX will not be able to be operated from amphibious carriers. The corps has 'borrowed' some MQ-8C Fire Scouts from the navy to test them on the big decks and evaluate what can be achieved by unmanned platforms in roles ranging from ISR to weapons delivery. This could well become another area of significant growth in the next decade.

The first *America* (LHA-6) class amphibious assault ship has spent much of the time since commissioning in October 2014 being modified to allow safe operations by F-35B Lightning II STOVL aircraft. This picture shows her conducting post-refit trials of her re-coated strengthened flight deck off the Californian coast in May 2016 with Harrier jets and Seahawk helicopters embarked. *(US Navy)*

Amphibious warfare capability in Australia took a big step forward with the commissioning into the Royal Australian Navy of their second new LHD type amphibious assault ship, *Adelaide*, in December 2015; the first LHD, *Canberra,* has already reached IOC. 808 NAS – equipped with MRH-90 Taipan multi-role helicopters – has been commissioned specifically to operate from them; the ships are capable of operating tailored air groups composed of a wide range of helicopters including Australian Army MRH-90s, Chinooks, Blackhawks and Tiger 'gunships' in addition to those fielded by 808 NAS. In the United Kingdom, the government announced after SDSR 2015 that the amphibious helicopter carrier *Ocean* is to be withdrawn from service, as originally planned, by 2019 with no direct replacement. However, at least one of the *Queen Elizabeth* class aircraft carriers is to be modified to operate in an amphibious role when required.

By 2017, the Egyptian Navy will have taken possession of its own amphibious assault ships; the two *Mistral* class ships originally built in France for the Russian Navy. A contract for their procurement was signed in October 2015 after which Russian specialists removed sensitive communications, combat systems and other equipment. It was subsequently announced that the Egyptian Navy is to acquire Ka-52K Katran attack helicopters from Russia for operation from these ships, possibly including some of the airframes originally built for the Russian Navy. In February 2016 Egyptian naval personnel arrived in France to begin training and prepare for the delivery of the ships in June and September 2016 respectively. The first ship, *Gamal Abdel Nasser* (the former *Vladivostok*), was handed over on schedule on 2 June 2016. Each is capable of embarking sixteen helicopters, four amphibious landing craft and up to 450 troops and supporting equipment. The training package is reported to cover tactical as well as technical matters. While there can be no doubt that it will take time for the Egyptian Navy to absorb these capable vessels, they offer a considerable step forward in its capability.

The Royal Australian Navy's second *Juan Carlos I* type amphibious assault ship *Adelaide* pictured undergoing readiness training in Jervis Bay during April 2016. The class has transformed the RAN's amphibious capabilities but proposals that F-35B Joint Strike Fighters should be deployed on the ships were not supported in the country's recent defence white paper. *(Royal Australian Navy)*

SEA CONTROL HELICOPTERS

2016 has seen consolidation of the sea control helicopter forces operated by many navies, with developmental difficulties overcome and production or modernisation runs nearing completion. The NH90 Caiman has now reached full operational capability with the French Navy and the broader NH90 programme is making significant progress in others. For example, in February 2016 the first Belgian naval NH90 pilot achieved deck landing qualification when, accompanied by a Dutch instructor, he carried out ten successful landings on a Royal Netherlands Navy frigate. NH Industries has also recently undertaken the first flight of a 'navalised' TTH transport variant of the NH90. This is equipped with enhancements, such as navalised landing gear and an automatically folding main rotor, that are found on the naval NFH frigate version.

Also making progress is the Finmeccanica Helicopters (formerly AgustaWestland) AW-159 Wildcat programme. In March 2016, the Philippine

The Egyptian Navy has been undergoing a remarkable expansion in recent years, most notably through the acquisition of the two *Mistral* LHD-type amphibious assault ships whose sale to Russia had been blocked following that country's military intervention in Ukraine and the Crimea. These images show the re-named *Gamal Abdel Nasser* (formerly *Vladivostok*) undergoing sea trials prior to delivery on 2 June 2016. *(Bruno Huriet)*

Navy became the third customer for the type when it signed a contract, reportedly worth €100m (c. US$115m), for two helicopters and associated support.

In the United States, the MQ-8C Fire Scout UAV continues to carry out successful development trials and detachments are expected to be embarked in 2017. Meanwhile, in the area of manned helicopters, the imminent end of MH-60 production for the US Navy is beginning to focus its attention on the type of platform that will eventually be required to replace it. The Pentagon has sponsored a joint investigation into requirements for a new generation of vertical-lift platforms. However, the immediate focus has been on the competition between the Bell/Lockheed-Martin V-280 Valor tilt-rotor and the Sikorsky/Boeing coaxial SB-1 Defiant, which are both intended to meet US Army requirements. The US Navy wants a degree of commonality with new army types but feels their high speed is less important than economical hover and loiter capability. Nevertheless, the navy hopes to harvest some common architecture and systems. The timescale for procuring an MH-60 replacement depends on the findings of a helicopter master plan and service life assessment programme (SLAP), which is evaluating how long the existing naval helicopter fleet can be sustained. Any follow-on platform will likely have to achieve IOC between 2030 and 2040, therefore allowing time to evaluate the US Army's new designs and then embark on a naval programme.

MARITIME PATROL AIRCRAFT (MPA)

The British Government's 2015 SDSR decision to procure nine P-8 Poseidon aircraft was the most significant recent development in the area of MPA operations. P-8As have been in operational service with the US Navy since 2013 and ten VP maritime patrol squadrons are expected to have achieved operational status with the type by the end of 2017. The first unmanned patrol squadron, VUP-19, is also due to achieve IOC with the MQ-4C Triton UAV in 2017 with a second unit to follow.

The improved MQ-8C variant of the Fire Scout UAV – which uses a larger airframe than the current MQ-8B – has continued to carry out successful development trials. It will start to replace its earlier sibling in operational detachments during 2017. *(Northrop Grumman)*

P-8Is are also in service with the Indian Navy, whilst the Australian Government initially decided to order eight P-8As for the Royal Australian Air Force (RAAF) – with an option for four more – to replace the Service's aging fleet of AP-3C Orions. Like the US Navy, the RAAF will pair the P-8 with the unmanned MQ-4C Triton, with both types being based at RAAF Edinburgh in South Australia. The first RAAF P-8A is to be delivered in 2017 with IOC anticipated in 2021. IOC with the Triton – seven of which will be acquired – should be achieved shortly afterwards. The February 2016 Australian Defence White Paper emphasised maritime security and increased the number of planned P-8As from eight to fifteen. The extra seven aircraft will be built in two tranches to be delivered in the late 2020s. The acquisition of the seven Tritons was also confirmed.

The United Kingdom has lacked an MPA capability since 2008, when the Nimrod MR 2 was grounded. Development of the projected replacement, the Nimrod MRA 4, which had suffered lengthy delays and a crippling increase in cost, was terminated in the 2010 SDSR. Hopes that other assets could fulfil at least part of the role were not realised. The answer to a 2016 Parliamentary Question revealed that the United Kingdom had come to rely heavily on NATO allies to fill the gap. During 2015, twenty-one MPAs were deployed temporarily to RAF Lossiemouth in Scotland at the request of the British Government; of these, eleven came from the United States, five from France, three from Canada and two from Germany. No date has yet been announced for the first British P-8 to be delivered but it would be reasonable to expect the first aircraft to be flying by 2018. It is likely that they will have the same equipment fit as the US Navy and RAAF aircraft since any modifications would be both time-consuming and expensive. Crew skills have been kept alive by RAF aircrew who have flown with the US Navy and these will form the nucleus of a new squadron, which will be based at RAF Lossiemouth. Given the limited number being procured, it is likely that only a single squadron will be operated, with air and ground crew training carried out by the US Navy in order to make the most effective use of every British aircraft. Surprisingly, the Government has made no statement about potential Royal Navy involvement in P-8 operations.

PROSPECTS FOR BRITISH MARITIME AVIATION

SDSR 2015: In February 2016 the then British First Sea Lord, Admiral Sir George Zambellas, gave a speech at The Navy Club in London in which he gave his views on Britain's SDSR 2015. He and the Admiralty Board had met in November 2014 to define their vision of the future Royal Navy. Despite political and budgetary uncertainties, they bravely elected to press for a maritime focus within British defence based on three core capabilities of (i) continuous at-sea deterrence, (ii) carrier strike and (iii) amphibious readiness. The new government elected in May 2015 undertook to spend two percent of national income on defence and this gave financial depth to the Admiralty Board's aims which were, for the most part, delivered. Of interest, one of the highest priorities was the restoration of an MPA capability since, although they are to be operated by the RAF, they were accepted to be an important element of the United Kingdom's maritime capability.[7] The commitment to acquire P-8 Poseidon aircraft was, possibly, SDSR 2015's major outcome.

Indeed, media speculation about massive cuts proved unfounded and SDSR 2015 can fairly be said to have delivered stability after decades of retrenchment and reductions in force levels. The aircraft carrier *Queen Elizabeth* will be completed for contractor's sea trials around the end of 2016 and is to be handed over to the Royal Navy in 2017 to commence a two-year first-of-class trials programme. This will include three protracted periods in American waters with F-35Bs embarked. Her sister-ship *Prince of Wales* is structurally complete and is to be floated out during 2017. The Prime Minister, David Cameron, described the SDSR as delivering 'a bigger Navy with more ships' but the Chancellor of the Exchequer, George Osborne, was more specific when he said that 'we are going to step up the aircraft carrier punch of the UK with twenty-four F-35Bs on the decks of the carriers by 2023, when Great Britain will be second only to the USA in its ability to project power in this way'. It had already been announced that both carriers would be commissioned and operated to guarantee at least one continuously available at short notice. The Chancellor's statement appeared to have taken little account of developments in India and China but at least the SDSR measures will make the United Kingdom a more stable and reliable partner for its allies.

The US Navy's P-8A Poseidon maritime patrol aircraft is steadily taking over from the veteran P-3 Orion in the US Navy's maritime patrol (VP) squadrons and is attracting much interest overseas. Australia has increased planned purchases from eight to fifteen over the course of the year, whilst the United Kingdom has become the fourth country to specify the type. *(US Navy)*

SDSR 2015 also reaffirmed a commitment to procure 138 F-35s 'over the life of the project'. This number was originally announced by a different party in government over a decade ago, so confirmation of the full buy was welcome. The phrase 'over the life of the project' is interesting, however, since the F-35 is likely to remain in production for the US Armed Forces and other export customers into the 2030s and probably well beyond that. British production could, thus, extend over another twenty years in annual batches of around six aircraft tied-in with United States' lots or multi-year buys.

SDSR 2015 intends to deliver a stable force into what it terms Joint Force 2025, effectively a British expeditionary task force of all arms ready for immediate action anywhere at short notice. From a maritime aviation perspective, the main operational contribution would appear to be focused on:

- Two F-35B Lightning II squadrons.
- A single P-8 Poseidon squadron (nine aircraft).
- Four Merlin HM 2 squadrons.
- Two Merlin HC 4 squadrons
- Two Wildcat squadrons

The review was light on specific detail, unfortunately, and made little mention of the fact that RAF Chinook and Army Air Corps Apache helicopters are expected to form part of tailored air groups (TAGs) in carriers for specific operations and exercises (together with Royal Navy Merlins and Joint Force F-35Bs). The omission must lead to concerns that the increasing number of squadrons with more than one role cannot be in two places at once and – if committed to an operation ashore for any length of time – may leave carrier decks empty. If this were to happen, lack of embarked aircraft would make it difficult to keep ships' companies at the peak of effi-

A computer-generated image of the new F-35 Maintenance & Finish Facility at RAF Marham, which will support the F-35 force that will be jointly operated by the Royal Air Force and Royal Navy. SDSR 2015 accelerated planned purchase of the new aircraft to ensure that twenty-four could be used operationally from a *Queen Elizabeth* carrier by the mid-2020s. *(BAE Systems)*

ciency needed to maintain constant readiness for operations; it is not just pilots who need to maintain embarked skills, it is the whole ship's company. It is also not entirely clear whether the list of squadrons quoted relates only to operational units or also includes training formations.

The Fleet Air Arm Structure in 2017: In the meantime, there is still a considerable way to go before the force structure planned for Joint Force 2025 becomes a reality. This is made evident by the following assessment of key elements of the Fleet Air Arm and F-35B Joint Force as 2017 approaches.

Fast Jet Aviation: As of mid-2016, the United Kingdom had taken delivery of just four F-35Bs, three of which are 'orange-wired' development aircraft. These three aircraft serve with 17 Squadron RAF at Edwards Air Force Base (AFB) in California on operational testing and analysis tasks. Although an RAF squadron, around half of the unit's manpower is provided by the Royal Navy. The unit is scheduled to be the first to land F-35Bs on *Queen Elizabeth* during her trials programme in 2018. The fourth aircraft is allocated to VMFAT-501, the 'Warlords', at Marine Corps Air Station (MCAS) Beaufort in South Carolina. This has Royal Navy and RAF personnel attached to it and trains British as well as US Marine Corps pilots and ground crew.

The United Kingdom also currently has ten F-35Bs on order in batches of four and six, the latter ordered in late 2015 as part of US Lot 9. They will be used to form 617 Squadron RAF at MCAS Beaufort in late 2016. It will also draw its personnel from both the RAF and Royal Navy. It will spend two years working up in the United States before reaching IOC in 2018, when it is expected to move to the Joint Force Base Marham in the UK. It is planned to be the first operational squadron and will be capable of embarking in *Queen Elizabeth* from 2020. Further annual production batches will go to 809 Naval Air Squadron (NAS), which is expected to form at Beaufort in 2017 and move to Marham in 2019. Like the other squadrons, it will draw manpower from both the air force and navy. Eventually, all British F-35Bs are to be based at the Joint Force Base Marham except for 17 Squadron, which will remain in the United States at a permanent location that has yet to be announced.

736 NAS, equipped with Hawk T 1 aircraft, currently provides a focal point for fast-jet aviation within the Royal Navy command structure and gives continuation flying for pilots before they join the F-35 programme; it maintains detachments at Royal Naval Air Stations (RNAS) Culdrose and Yeovilton.

Sea King ASaC 7 and Merlin HM 2 Helicopters: The Royal Navy's airborne surveillance and control force has shrunk with the decommissioning of 854 and 857 Naval Air Squadrons but a core capability has been maintained with the retention of seven Sea King ASaC 7 helicopters in 849 NAS. These are the last operational Sea Kings in the Royal Navy, survivors of a fleet that first entered service in 1969. The squadron trains ASaC observers, provides a focus for tactical development and can deploy two flights, named 'Normandy' and 'Palembang' (after two of the squadron's battle honours) as part of a TAG until late 2018.

From 2019 ASaC capability is to be provided by 'Crowsnest' packages, comprising an upgraded version of the Searchwater 2000 radar and Cerberus mission system fitted in the Sea King. All thirty RN Merlin HM 2s are to be modified to allow 'Crowsnest' to replace their sonar systems and associated control consoles. Ten packages are to be procured, with up to eight fitted in Merlins at any one time. They comprise a pallet of avionics tailored to fit inside the Merlin fuselage and a specially-designed retractable radome that attaches to one of the weapon pylons. The 'Crowsnest' concept was hailed by the British Ministry of Defence as providing greater flexibility for the Merlin force when it was announced, since it allows ten aircraft to 'swing' between roles with equipment changes that can be carried out on board. However, the need to maintain aircrew and carrier proficiency in the ASaC role suggests that helicopters will be assigned to this role semi-permanently, reducing the number of anti-submarine Merlins by an equivalent number. Twelve Merlins were not modified to HM 2 standard and it would have made more sense to use them as ASaC assets.

Four units currently operate Merlin HM 2s, which are based ashore at RNAS Culdrose. One of these, 824 NAS, is a training and tactical development unit. Another, 829 NAS, deploys Merlins to specialised anti-submarine frigates and the remaining two, 814 and 820 NAS, deploy aircraft to TAGs as necessary. It has yet to be made clear how evolution towards Joint Force 2025 and the incorpo-

Right: The Royal Navy's fleet of anti-submarine helicopters is focused on the Finmeccanica Helicopters (the rebranded AgustaWestland) Merlin, thirty of which have been upgraded to the latest HM2 standard. These aircraft will also carry out airborne surveillance and control tasking from 2019, using 'Crowsnest' radar and mission system packages. There are arguably insufficient of them for both tasks. *(Crown Copyright 2014)*

Below: Although most media attention has focused on the lead ship in the *Queen Elizabeth* carrier programme, which should be ready to commence trials around the end of 2016, progress on her sister-ship, *Prince of Wales*, has also been dramatic. This image shows her rear island structure being lowered into place in January 2015. SDSR 2015 confirmed that both ships of the class would enter operational service. *(Aircraft Carrier Alliance)*

ration of 849 NAS will alter this structure. For example, two or more Merlin squadrons could be amalgamated to form a headquarters from which TAG deployments will be made.

Transport Helicopters: The last Sea King HC 4 helicopters were retired from the Commando Helicopter Force (CHF), in March 2016 having given outstanding service since 1981. This left only seven Merlin HC 3(interim) helicopters capable of embarked operations. They are fitted with power-folding main rotors removed from Merlin HM 1s in storage, together with strengthened undercarriages and lashing points to allow safe deck operation. They are operated by 845 and 846 NAS based ashore at RNAS Yeovilton together with a variable number of un-modified Merlin HC 3s. Work has started at Finmeccanica Helicopters (formerly AgustaWestland) to convert the whole fleet of twenty-five aircraft to HC 4 standard with power-folding rotors and tail pylons, strengthened undercarriages, upgraded communications, improved defensive aids and a 'glass cockpit' similar to the Merlin HM 2. However, the last HC 4 is not due to be delivered until 2022 so the CHF will have to operate with limited capability until then. Once fully equipped the two squadrons will each have ten Merlin HC 4s, with the balance of five cycled through the front-line to replace aircraft lost, damaged or in deep maintenance. 845 NAS is to be capable of deploying three operational flights; 846 NAS is to have an operational conversion flight, a maritime counter-terrorism flight and an operational flight capable of backing up 845 but, given the small number of aircraft involved, the flights will not be large. A third CHF squadron, 847 NAS, operates the Wildcat AH 1 in the reconnaissance and light strike role to support 3 Commando Brigade.

The Royal Navy's force of transport helicopters is steadily transitioning towards a fleet of modified Merlin HC 3s previously operated by the RAF following the final withdrawal of the last Sea King HC 4 helicopters serving with the Commando Helicopter Force (CHF) in March 2016. The work required to modify CHF is quite substantial and it will not be until 2022 that work to upgrade all to the full HC 4 naval standard will be completed. In the meantime, some un-modified machines and a few upgraded to an interim HC 3(i) naval standard are providing a more limited capability. The images show a Sea King from 845 NAS operating in Norway in 2014 and Merlins from 846 NAS on a similar deployment the following year. *(Crown Copyright 2014, 2015)*

Lynx and Wildcat: March 2017 will see the retirement of the Lynx HMA 8 which has been in service since 1977 and completion of its replacement by the Wildcat HMA 2. It is operated by two squadrons, a training and development unit, 825 NAS, and an operational unit, 815 NAS, which deploys flights to destroyers and to frigates that do not operate Merlins. Both are shore-based at RNAS Yeovilton. Despite external similarities, the Wildcat is not a Lynx upgrade but an entirely new, somewhat larger and heavier, aircraft built around a digital avionics management system designed to enhance mission effectiveness. Unfortunately delays in procuring the Martlet and Sea Venom air-to-surface missile systems for this aircraft mean that it will not achieve its designed strike potential until 2020 or later. That said, the Wildcat is proving successful and reliable in service.

Conclusion: The 'end-game' of British naval aviation is to have a strike carrier task force capable of a wide spectrum of operations available at all times as the core element of the nation's response capability, and to provide every warship with at least one aircraft. IOC for the first carrier – with some F-35Bs embarked – is planned for 2020. However, the full capability – with a significant number of strike fighters equipped with a wide range of British and American weapons, 'Crowsnest' and the ability to support a flexible range of TAGs – may not be reached until 2023. The carrier TAG capability spectrum ranges from the embarkation of up to thirty-six F-35s with both versions of the Merlin HM 2 for a strike-focused deployment to a littoral manoeuvre configuration with up to 250 Royal Marines embarked in dedicated accommodation together with RAF Chinooks, Royal Navy Merlin HC 4s, Wildcat AH 1 and Army Air Corps Apaches. Somewhere between the two extremes is an expeditionary strike capability with a mixture of aircraft types embarked with supporting marines and, potentially, elements of a US Marine Corps air/ground task force.

The new generation of embarked aircraft all rely on software-controlled, digital management systems for maintenance, logistic support, training, personnel qualification and flight planning. Compatibility with other F-35B users and the global supply chain is to be achieved through ALIS. However, the system does not make full use of existing British supply management systems or those used by other aircraft. The Merlin HM 2 and Wildcat also have digital management systems, the latter shared with the Army Air Corps, but neither has interface with ALIS. This means that the navy is using entirely different software-management systems for aircraft, each created with the laudable aim of increasing commonality with other operators but none of which are common across the Royal Navy air engineering discipline. The implications for training and career development are far-reaching. With the simple, analogue and paper-based systems of the past, it was possible for maintenance personnel to move from one aircraft type to another using common engineering techniques but this can no longer be as straightforward. It will be interesting to see to what extent sailors become 'locked' into a single type once they have been trained and overcome the complexity of its constantly-evolving software systems. The F-35B, for example, is certainly not just a new aeroplane, but part of a very complex system of systems that will produce remarkable results and form a key element of our ability to fight the digital wars of the future. However, we are still a long way from fully understanding what the aircraft can deliver on a regular, reliable basis and what it will take to achieve it.

Notes

1. The actual situation obviously changes on an almost continuous basis. As of early 2016, *Harry S. Truman* (CVN-75) and *John C. Stennis* (CVN-74) were deployed, with *Dwight D. Eisenhower* (CVN-69) preparing for deployment. A further two ships, *Carl Vinson* (CVN-70) and *Theodore Roosevelt* (CVN-71) had recently completed deployments whilst another four ships were in – or preparing for – various stages of maintenance in the United States. In addition, *Ronald Reagan* (CVN-76), which is permanently forward-based in Japan, was also undergoing planned maintenance. *Gerald R. Ford* (CVN-78) had yet to be completed. Later, in June, 2016, Christopher Cavas of *Defense News* reported four carriers were deployed, the first time this had happened since November 2012. See 'US Navy Deploys Most Carrier Strike Groups Since 2012', *Defense News* – 6 June 2016 (Springfield, VA: Sightline Media Group, 2016).

2. See Seth Cropsey, Bryan G McGrath and Timothy A Walton, *Sharpening the Spear: The Carrier, the Joint Force, and High-End Conflict* (Washington DC: Hudson Institute, 2015).

3. At the end of the Cold War, the US Navy had to decide whether to develop a new long-range attack aircraft for the deep strike role or to focus funds on the development of long-range missiles. It was decided that long-range strikes would generally be against discrete targets, with mass attacks by naval aircraft more usually taking place at shorter ranges in the littoral. As such, it was determined the relatively short-ranged F/A-18 series of Hornet/Super Hornet strike fighter would be sufficient.

4. *Vikrant* takes the name of India's first aircraft carrier, a former British *Majestic* class vessel commissioned in 1961 and withdrawn from service in 1997 before being finally scrapped in 2014. Given considerable delays with Indian shipbuilding projects, the new *Vikrant*'s 2018 planned commissioning date must be regarded as speculative.

5. These numbers, contained in a US Marine Corps news release on 12 August 2015, are slightly different from previously quoted figures, which suggested a total USMC procurement of 340 F-35Bs and eighty F-35Cs.

6. A copy of the full report can be found at: http://www.dote.osd.mil/pub/reports/FY2015/

7. It should also be noted that the British F-35B Lightning II Force is not entirely part of the Royal Navy's Fleet Air Arm but nor is it entirely part of the RAF; it is being built up as a joint force in which both services have a stake. It is expected to be a 'transformational' force and the ideas discussed above about the way the USMC hopes to operate the F-35B apply equally to the British force. It can also be expected that British and USMC units will be seen working closely together in the years ahead.

8. This chapter has been compiled from a wide range of periodicals, of which *Air International, Flight, Jane's Defence Weekly, Warship World* and *The Navy* (the journal of the Navy League of Australia) provide particularly good sources of further reading. Reference should also be made to the following publications, as well as to the websites of relevant aircraft manufacturers and navies:

- Gunter Endres and Michael J Gething, *Janes's Aircraft Recognition Guide- Fifth Edition* (London: Collins-Jane's, 2007)
- Norman Friedman, *The Naval Institute Guide to World Naval Weapon Systems – Fifth Edition* (Annapolis, MD: Naval Institute Press, 2006)
- *Jane's Fighting Ships* – Various Editions (Coulsdon, Surrey: IHS Jane's)

4.2 TECHNOLOGICAL REVIEW

Author:
Norman Friedman

SHIPBOARD ANTI-SHIP MISSILES

SOME CONSIDERATIONS

In 2016 there seems to be a resurgence of interest, at least in the United States, in shipboard anti-ship missiles. The last major US Navy developments came during the latter part of the Cold War, with improvements to Harpoon and the development of the anti-ship version of the much longer-range Tomahawk. Now the US Department of Defense has announced an anti-ship version of the SM-6 anti-aircraft missile, work is proceeding on an anti-ship version of Tomahawk Block IV and several companies are offering new missiles, including a ship-launched version of the Norwegian Naval Strike Missile (NSM). To a large extent renewed American interest reflects a strategic shift, away from the littoral emphasis of the post-Cold War period and back towards a possible peer-competitor confrontation, this time in Asian waters against the growing Chinese navy. On a subtler level, renewed interest in anti-ship missiles also reflects changing ocean surveillance capabilities.

What follows is an outline of some considerations in the design and use of ship-launched anti-ship missiles; it is by no means a history nor a comprehensive survey of current and projected systems.

The American Harpoon – seen here being launched from the cruiser *Shiloh* (CG-67) during an exercise in 2014 – has probably been the most popular Western anti-ship missile. It exemplifies a philosophy which has been followed since the 1970s: a focus on stealth, achieved largely by flying at sea-skimming altitude. To achieve stealth, missile designers in the West have accepted subsonic speed but this may not be sufficient to penetrate modern defensive systems. As such, there is considerable work underway in the United States to develop improved shipboard anti-surface missiles. *(US Navy)*

COLD WAR DEVELOPMENTS

The fundamental problem a ship-launched anti-ship missile system must overcome lies not in the technology of the missile itself but in the supporting ability to find and track ships beyond the horizon of the firing ship. That is why the early combat use of such missiles, notably off Egypt in 1967 and then during the Yom Kippur War in 1973 and during a fight between Chinese and South Vietnamese naval forces the following year, was conducted at very limited ranges. In the first engagement, Egyptian missile boats inside the harbour at Alexandria used P-15 Termit (NATO: SS-N-2 'Styx') missiles to sink the Israeli destroyer *Eilat* patrolling offshore. It did not matter whether the missile boats could actually see the destroyer; her movements were entirely predictable. During the larger missile battles in 1973, both sides fired at targets visible on their radars; the Israeli boats were also probably supported by shore radar and shore control. The 1974 engagement in the South China Sea was fought at short range, its most notable lesson being the rather limited lethality of a 'Styx' missile which hit a South Vietnamese frigate (a former US Navy *Barnegat* class small seaplane tender). Unlike the *Eilat*, of similar size, this ship did not sink, although photographs show a large hole in her bow.[1]

When the US Navy bought ship-launched Harpoons for its surface combatants, it also bought LAMPS helicopters, whose multiple roles – indicated by the M in the designation – included over-the-horizon radar surveillance, the radar data being fed directly into the ship's combat direction system. Without such surveillance, the US Navy assumed that despite its nominal range (at that time) of 65nm (121km), Harpoon would not really be usable much beyond 14nm (26km).

The Cold War Soviet navy recognised the problem and developed two solutions. One was the use of a radar aircraft, generally a 'Bear D' (Tu-95RTs), which could provide an over-the-horizon radar picture to support long-range anti-ship missiles, such as the SS-N-3, SS-N-12 and SS-N-19, fired by ships and submarines. This solution was well understood in the West, as it corresponded to Western use of airborne early warning (AEW) aircraft in surface warfare. It was clear that an attack might well be aborted by denying the enemy access to this information and, hence, that the relatively low-performance 'Bear' was well worth shooting down. The Royal Navy armed Sea Harrier STOVL jets with Sidewinder air-to-air missiles for this reason. The Soviets were also well aware of how important – and potentially vulnerable – the 'Bear Ds' were. They tried to develop satellite systems to do much the same job but the key element of the system, a radar satellite powered by a small reactor, failed.

The other Soviet solution was unknown to the West during the Cold War; its details emerged only

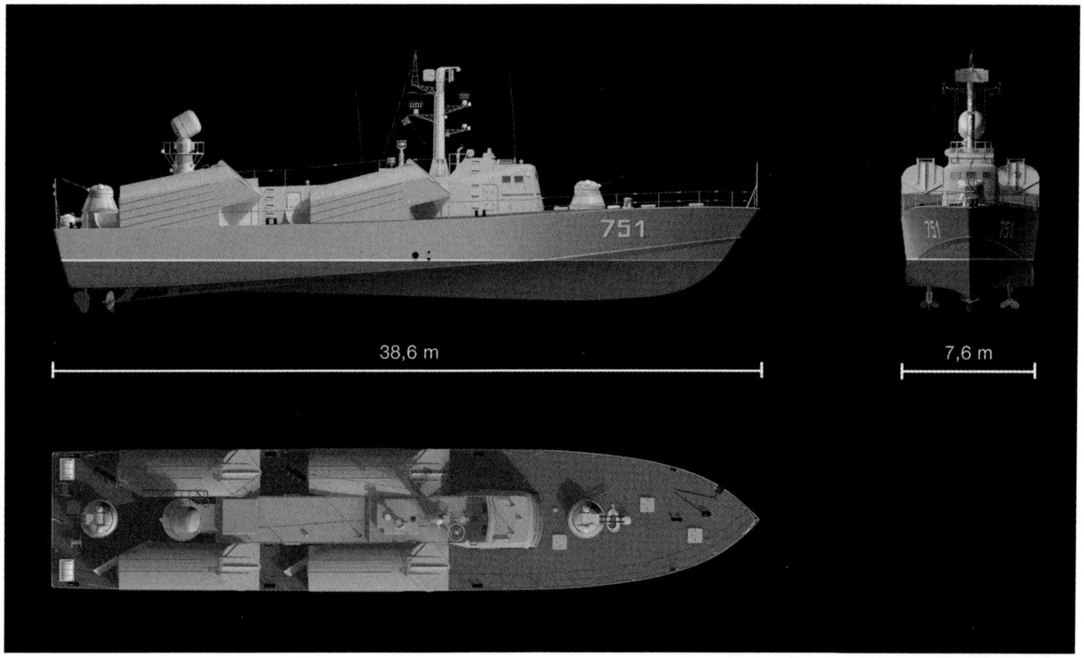

A graphic of the Russian-designed Project 205 'Osa I' type fast attack craft, which was created as a platform for the SS-N-2 'Styx' missile. The four launchers for the missile are clearly visible. A large weapon, the 'Styx' was responsible for sinking the Israeli destroyer *Eilat* in 1967 and was used successfully by India during raids on the port of Karachi during the 1971 Indo-Pakistani War. *(Max Eberhard)*

A Russian Tu-95 'Bear H' strategic bomber pictured in 2014. The Tu-95RT 'Bear D' variant of the aircraft – used to provide an over-the-horizon radar picture for Soviet warships – was deployed as one solution to the problem of designating targets for long-range anti-ship missiles *(Crown Copyright 2014)*

An aerial view of the forward part of the Russian Project 956 *Sovremenny* class destroyer *Otlichnnyy*. In addition to deploying radar reconnaissance aircraft, the Soviet Union also found a way to detect ship targets beyond a firing ship's horizon by using two complementary techniques: troposcatter and ducting. The big 'Band Stand' radome in front of the bridge of this *Sovremenny* class destroyer houses antennas for both. The big canisters contain eight P-270 Moskit (SS-N-22 'Sunburn') missiles. *(US Department of Defense)*

A prototype Tomahawk Block IV cruise missile – also initially known as a Tactical Tomahawk – pictured during flight testing in 2002. The US Navy developed anti-ship variants of the Tomahawk land-attack cruise missile during the Cold War but these were abandoned after its end, partly due to targeting difficulties at long ranges. The navy has subsequently demonstrated that a net based on periodic GPS updates of target positions can transmit enough data to a missile to guide it into a moving ship even if the missile has no onboard seeker. In practice, however, the new anti-ship variant of Tomahawk now being developed will incorporate a seeker. *(US Navy)*

later, when Russians wrote about their Cold War achievements. It was a shipboard over-the-horizon radar, part of a system the Russians eventually called Mineral.[2] The radar antennas occupy the radome NATO called 'Band Stand', which initially appeared on board 'Nanuchka' class missile corvettes, supporting the P-120 Malakhit (SS-N-9 'Siren') anti-ship missile. Later, 'Band Stand' appeared on board *Sovremenny* class missile destroyers, supporting the SS-N-22 system. In this form it was exported to China, and the Chinese installed it on board some of their own destroyers.

'Band Stand' encloses two antennas mounted back to back. One of them employs a troposcatter technique. This is commonly used for communication but not for radar, as too much of a signal is lost on the way (communication uses a one-way path, but a radar needs the return path as well). The signal scatters off the troposphere, a layer about 50 miles (80km) up in the atmosphere. The other antenna is used to detect radar signals which pass over the horizon due to ducting (anomalous propagation). Ducting is common in many areas the Soviets frequented, such as the Baltic, the Eastern Mediterranean and the South China Sea. The two antennas operate at different frequencies for which the two propagation modes are most effective: L Band (NATO C/D Band) for troposcatter and X Band (I/J Band) for ducting. The nature of 'Band Stand' was probably a surprise because during the Cold War the principal Western navies considered over-the-horizon radar propagation – at least from ships – as a nuisance (causing confusion) rather than as a positive asset. Apparently only the Royal Swedish Navy, operating in the Baltic, actively exploited the ducting it generally experienced, seeing it as a method of achieving useful surveillance ranges for its missile boats.

By the early 1970s the US Navy was interested in acquiring its own anti-surface missiles as a means of countering Soviet ships and submarines armed with long-range anti-ship missiles. The submarine force proposed a long-range supersonic anti-ship missile, to be carried on board a new, ultimately abortive class of submarines. The new generation of low-frequency array sonars offered the possibility of detection beyond (actually below) the horizon.[3] The initial proposal also envisaged information fed to the submarines by other platforms via a new satellite-based information distribution system. This particular missile was abandoned together with the huge submarine with which it was associated. Instead development of a submarine-launched version of Harpoon, at that time an air-launched anti-ship missile, was ordered. So was a ship-launched Harpoon. Once the navy began developing the much longer-range Tomahawk strategic cruise missile, an anti-ship version of Tomahawk was added, for launch by submarines and surface ships.

None of these missiles presented enormous technical problems. By the early 1970s radar and missile guidance technology were good enough to bring a missile to a designated point and to lock it onto a ship target it might detect. It soon emerged that the real problem was for a ship or submarine tens or hundreds of miles from such a ship to detect it, and, moreover, to determine which ship in a formation was the one to attack. The US Navy was certainly aware of the problem from the defensive point of view, as it used blip-enhancing jammers intended to prevent a Soviet radar from distinguishing the carrier in a formation. The Soviets were also aware of the problem; they deployed surface ship 'tattletales' to steam with Western carrier groups specifically to call out which was the right target.[4]

This was an ocean surveillance problem: the firing ship or authority needed to know where Soviet ships were within range of the shooter. For a time the US Navy's preferred solution was the same as the Soviet one: a dedicated radar ocean surveillance satellite. The satellite project died in the mid-1970s because it was unaffordable, leaving what seemed to be a hole in American capability. The solution adopted at the time was ingenious. A US officer observed that US intelligence already collected so much information that it might well be possible to track Soviet ships by collating what was already available, rather than by buying a dedicated system. For example, the satellites collecting Soviet communications traffic also collected traces of radar emissions, which were typically discarded. Instead, they could be exploited. Such a satellite would locate a ship on each pass. Major Soviet radars were hand-made, so that emissions by each particular radar on board each ship could be distinguished.[5] In this way the global pattern of Soviet naval deployments could be tracked, and a ship with a 250nm (464km) missile could certainly fire into a group of Soviet warships with reasonable confidence of hitting one.

There was a rub. Most ships at sea were not Soviet warships, so a missile might well lock onto a freighter or a neutral warship rather than the intended target. Critics of the passive satellite fleet-tracking system pointed out that most missiles would hit neutrals. Advocates pointed out that the system was entirely passive, so it gave the Soviets no warning whatever that they were being tracked. Missile strikes would be surprises. Even if many of the missiles were wasted, the effect of surprise on the Soviet fleet would be enormous. To help an approaching Tomahawk select the right target, anti-ship Tomahawks were fitted with a radar receiver which could recognize the Head Net C auxiliary height-finder on board the two of the most important Soviet warships, the two *Moskva* class helicopter carriers. This was a very imaginative solution but, unfortunately, the receiver was hard-wired (i.e., not reprogrammable), and this particular radar was obsolescent. It is not surprising that the anti-ship Tomahawk was retired as the Cold War ended.

THE POST-COLD WAR ERA

After the end of the Cold War, the main Western naval role changed dramatically. Instead of open-ocean warfare against the Soviet fleet, the Western navies expected to fight coastal wars against minor powers. These might have fast attack craft armed with anti-ship missiles. The principal Western anti-ship missiles, such as Harpoon, were modified so that they could be directed against fixed shore targets and also against ships near coasts. For the former,

The French Exocet, which has been very popular over a considerable timescale, was conceived for fast attack craft with limited ability to detect targets. Its designers considered a short-range rocket sufficient. However, detection ranges have increased, so Exocet has now been produced in the turbojet-powered Block III version shown in this schematic and image of a test firing. *(MBDA/Michel Hans)*

The Chinese Type 052 destroyer *Harbin* pictured in Sydney Harbour during a goodwill visit to Australia in September 2007. Her battery of sixteen C-802 surface-to-surface missiles can be seen amidships. Chinese ships tend to deploy such missiles in greater numbers than their Western counterparts and their ships have benefitted considerably from Western technology. *(Royal Australian Navy)*

The Indian modified Project 65 *Rajput* class destroyer *Ranvijay* pictured operating with the US Navy in 2012. The ship is noteworthy in deploying both one of the earliest (albeit upgraded) generation of shipboard anti-ship missile systems in the form of the SS-N-2D 'Styx' series located in box launchers to either side of the forward superstructure and the latest Indo-Russian supersonic BrahMos missile, located in an eight-cell launcher aft the second pair of funnels. BrahMos is reportedly capable of speeds of up to Mach 2.8, continuing the Soviet approach of using high speed to leave defensive systems little reaction time. *(US Navy)*

they were given auxiliary GPS guidance, which would direct them as desired. For the latter, their radar seekers were upgraded so that they could distinguish targets close to shore or in a harbour's confines. Generally this meant Doppler radar.

There was a problem to this approach. Western warships, typically, were armed with four or eight anti-ship missiles installed in canisters, which are hard or impossible to replenish at sea. It is difficult to see eight 250lb (113kg) warheads as decisive shore bombardment weapons. At present there is no vertically-launched anti-ship missile in US Navy service. The new anti-ship Tomahawk and SM-6 will be vertically launched but, even then, the numbers deployed on any one ship will generally be very limited. Vertical launchers also generally cannot be replenished at sea.

During the first decade of the new millennium, the most important 'new' Western anti-ship missile was the updated Exocet MM40 Block 3, in which the short-range solid-fuel motor of the earlier versions was replaced by a turbojet. The missile was also redesigned to make it stealthier. For example, it now has a radome which passes only a limited range of radar frequencies. As such, a radar observing it does not normally see the missile's flat radar seeker but, instead, sees the stealthier shape of the dome. Block 3 is also adapted for shore bombardment. More recently the French Navy has developed a cruise missile broadly comparable to Tomahawk called Scalp Naval and it will presumably have an anti-ship capability.

Meanwhile the Chinese have been fielding longer-range anti-ship missiles of their own on board destroyers equipped with the Russian Mineral system or its Chinese derivative. They have also been deploying those missiles in greater numbers. Thus a Type 052B destroyer, for example, carries sixteen missiles in canister launchers, compared to the more usual eight in the West.

The US Navy has indicated its own interest in a next-generation anti-ship missile, and considerable attention is being paid to supersonic and hypersonic technology. The argument is that, although subsonic missiles like Harpoon were effective against earlier air defences, it takes something more to penetrate modern systems. A supporting argument is that tactical data are fleeting, so that the sooner the missile reaches the last reported (or projected) target position, the better the chance that it will find a target there. The Russians have made both these

The US Navy is looking to modernise its inventory of anti-ship missiles, with the ultimate aim being to field an advanced anti-ship missile in the 2020s. In the interim, work is underway to field the interim LRASM (Long Range Anti-Ship Missile) for airborne use, whilst the Norwegian NSM (Naval Strike Missile) has been fired from an improvised launcher on a Littoral Combat Ship. These images show LRASM on a F/A-18E Super Hornet and the NSM being test-fired from the Norwegian fast attack craft *Gnist. (US Navy/Norwegian Armed Forces)*

arguments when advertising their own current supersonic weapons. They are P-800 Oniks/Yakhont (SS-N-26 'Strobile') – which is closely related to the Russian-Indian Brahmos – and the Novator P-900 Alfa. The former is a supersonic rocket-ramjet. The latter is a version of the Russian naval land-attack cruise missile with a supersonic terminal stage. The Russians have also been selling the subsonic Uran (SS-N-25 'Switchblade'), which resembles Harpoon, and the subsonic 3M54 Kalibr/Club series, which resembles Tomahawk.

The US Navy's long-term plans focus on a future Advanced Anti-Ship Missile, which is to be acquired under the Offensive Anti-Surface Warfare/Increment 2 weapon programme to enter service in the mid-2020s. This will be launched from either a vertical launcher on board a ship or from a plane. There are a number of potential contenders for this programme, including Lockheed Martin's Long Range Anti-Ship Missile (LRASM), which has already been adopted for airborne use under designation AGM-158C. In the meantime, a modified Tomahawk Block IV with an active anti-ship seeker has been put into development and a special Department of Defense organisation developing new roles for existing weapons has also initiated work on an anti-ship derivative of the current SM-6 anti-aircraft missile. This exploits that weapon's programmable active seeker and its commandable autopilot, with GPS input. There is also considerable interest in the Norwegian Naval Strike Missile which Raytheon is promoting in partnership with Kongsberg. The missile was successfully tested from an improvised launcher mounted on the deck of the littoral combat ship *Coronado* (LCS-4) in September 2014.

CURRENT CONSIDERATIONS

Has anything changed since long-range Tomahawk was a questionable idea? The issue is really still whether the command system which fires the missile has some way of knowing what the situation is well beyond the horizon. The sea is still vast and consistent ocean surveillance is difficult at best. A surface warship on its own has access to a somewhat better ocean surveillance picture than in the past, but ships can still evade surveillance by shutting down their emitters. Unless the missile is smart enough to distinguish a warship from other targets, a ship can still evade attack – if it is aware that it is under threat – by closing with a much larger radar target, a merchant ship which cannot outrun it. A radar-guided missile, which turns on its radar near a mixed group of warships and merchant ships, is still likeliest to lock onto a merchant ship, particularly if the signature of the warship has been reduced by stealthy design. There is, of course, the possibility of exploiting the target's emissions. However, many modern radars and command links are much

The development of long-endurance unmanned aerial vehicles (UAVs) such as Northrop Grumman MQ-4C Triton offer the possibility of replicating the over-the-horizon radar reconnaissance role carried out during the Cold War by the Soviet 'Bear'. However, such UAVs would probably be even more vulnerable in hostile conditions. *(Northrop Grumman)*

Current anti-ship missiles are capable of disabling or significantly damaging small vessels but have only infrequently destroyed their targets. An exception was the Iranian frigate *Sahand* (top), although it took three air-launched Harpoon missiles and a number of cluster munitions to cause this level of destruction. Conversely, the US Navy frigate *Stark* (FFG-31) (above) survived the effects of two Exocet missiles. *(US Navy)*

stealthier than their predecessors, even though there are still characteristic ones. On the other hand, many modern surface ships have long-range passive sonar – generally meaning a towed array – and this may provide a fairly distant detection horizon. Unfortunately for the US Navy, the South China Sea is generally described as a very bad sonar environment, and at present Chinese warships there have a considerable advantage in the form of their version of Mineral/'Band Stand'.

To the extent that there is a difference between the situation now and in the past, it is provided by long-range long-endurance unmanned platforms. These can be used as snoopers, much as the Cold War Soviets used their 'Bear Ds'. A big unmanned aerial vehicle (UAV) can provide a sufficient radar picture to enable targeting and it can photograph individual ships so that they can be identified. Compared to a 'Bear D', UAVs offer lower performance – though perhaps higher altitude – and less powerful sensors, particularly radars. There is no one on board to detect hostile air activity and to order evasion, and it is unlikely that the intelligence resident in the UAV makes up that difference. It is difficult to imagine something like a Global Hawk or its Chinese equivalent surviving for long under hostile conditions. On the other hand, a surface ship within a fleet can exploit over-the-horizon radar coverage such as that provided by an E-2D, which would presumably be protected by fleet firepower. The US Navy is currently emphasising the ability to link the sensors of a battle group to create a truly joint picture on the basis of which weapons can be fired. This new capability probably justifies a new generation of anti-ship missiles. Even if the shipboard versions are not really any more usable than the Cold War Tomahawk, they have air-launched versions which can exploit the far better picture offered by airborne radars and other sensors.

One other point is worth raising: lethality. When the Israeli destroyer *Eilat* was sunk in 1967, and again when the British destroyer *Sheffield* was destroyed by an air-launched Exocet missile during the 1982 Falklands War, there was a widespread perception that surface ships were finished. Homing missiles could – and would – find their targets, and a single hit would, at the least, disable and, probably, sink a ship. In the case of *Sheffield*, it now seems that the damage was done by a massive oil fire. Initially it was believed that the Exocet had not exploded, hence that the fire was ignited by a motor still burning, although now it seems possible that the missile did indeed explode, but that its small warhead was not noticed.[6] The survival of US Navy frigate *Stark* (FFG-31) under worse conditions suggests that such missiles are not nearly as lethal as had been imagined. Reportedly, some of the results of tests used to demonstrate various missiles may even have been improved by placing explosives in strategic positions on board the targets.

The argument has been that all of this is irrelevant: a hit on a modern warship entirely dependent on electronic systems would be so disabling as to be fatal. However, over time modern electronics have grown more and more rugged, as any user of home electronics can testify. It is also possible to distribute and duplicate functions so that no single hit (possibly no two or three hits) will be fatal. Such dispersal might make it possible to reduce the scale and cost of active defence, which at present is required to destroy all incoming weapons. At the very least it should change perceptions of the future value of anti-ship missiles.

In this contest it is remarkable that no current missile is designed to achieve an area effect on a ship. If the ship can withstand point attacks, then the best target – the one a torpedo strikes – is the hull girder as a whole. A torpedo exploded under a ship's keel can break it in two. A ship can be designed to survive that sort of damage, but that requires a radically different approach to design and much greater ship size. In the 1950s the Soviets deployed an anti-ship missile designed to drop its warhead into the water alongside a ship to achieve exactly this effect, but it proved ineffective. Would a more modern equivalent work? Are there other mechanisms likely to sink a ship?

If this seems very academic, consider the problem from the missile shooter's point of view. He fires a salvo of missiles. They may or may not survive enemy anti-missile fire, and they may or may not work at all. Once they pass over the horizon, those questions are difficult or impossible to answer. If electronic emissions and sounds from the target ship cease, does that mean that it has been disabled, or does it mean that a very cold-blooded enemy commander is playing possum? The objective of actually sinking ships is often decried as unnecessarily expensive (or excessively aggressive), but a ship which is sunk or blown in half offers a clear measure of damage assessment. Nothing else really does.

Notes

1. 'Styx' missiles were also used to good effect during Indian raids on Karachi in 1971 during the Indo-Pakistani War. Aside from the early incidents, attacks by anti-ship missiles have generally been by aircraft, famous cases being the sinking of the British destroyer *Sheffield* and the container ship *Atlantic Conveyor* in the Falklands in 1982 and the near-fatal Exocet attack against the US Navy's *Stark* (FFG-31) in 1987. Air-launched Exocet missiles were used extensively by Iraq during the 'tanker war' which accompanied the Iran-Iraq War, and lighter, helicopter-launched Sea Skua missiles were used effectively against Iraqi fast attack craft during the 1991 Gulf War. The Hezbollah attack on the Israeli Navy corvette *Hanit* off Lebanon in 2006 by a Chinese-designed C-802 missile was both unusual and largely ineffective.

2. Five different systems used the same radome. The first was Dubrava, in 'Nanuchka' class missile corvettes, controlling the P-120 Malakhit (SS-N-9 'Siren') missile. Clearly Dubrava was sensitive technology: export 'Nanuchka' class vessels had a conventional 'Square Tie' radar in the 'Band Stand' radome. Dubrava was eventually replaced by a Titanit system, whilst 'Tarantul' class missile boats and the prototype Project 956 missile destroyer *Sovremenny* had the follow-on Monolit system. Later *Sovremenny* class ships, including those exported to China, had Mineral. Titanit was the first in the series to combine independent active and passive channels, a reception channel for an airborne radar, a means of processing and displaying the data, a data link between ships of a tactical group, and a navigational radar. The associated 'Light Bulb' radome contained the antenna for the directional data link (some Russian accounts associate 'Light Bulb' with a passive channel, presumably ESM). In the Dubrava system, the 'Fish Bowl' antenna was the associated data link. Russian brochures give Mineral active range as up to 97nm (180km) and passive range as up to 243nm (450km). Claimed target capacity was fifty targets tracked passively and thirty actively, with 200 carried by the data link.

3. Moreover, the passive sonars might identify a potential target by its acoustic signature. Much the same could be said of long-range surface ship sonars operating in passive mode, and then of the longer-range towed arrays. Detection of distant surface targets by such sonars was generally called 'below the horizon'. Ranges might be two or even three convergence zones under good conditions, which would mean 70 or 105nm (130 or 195km) in the Atlantic, and 40 or 60nm (74 to 111km) in the Mediterranean. However, passive sonar ranges and bearings were never very precise; at 70nm a bearing within 15° (which was considered good) left open a distance of about 18nm (33km) A convergence zone would be 5 to 10nm (9 to 18.6km) wide. Manoeuvring and analysis of how a target seemed to move could reduce these figures somewhat.

4. The tattletales were destroyers – hence fast enough to keep up with the targets – which could be distinguished by their aft-facing batteries of short-range P-15 Termit (SS-N-2 'Styx') missiles. Their role seems not to have been obvious in the West, at least initially. Many thought of them as the primary means by which the Soviets tracked the carriers. That was not the case; the Soviets relied mainly on big HF/DF arrays in the Soviet Union. The aft-firing missile batteries made perfect sense: when an attack was imminent, the tattletale would turn back to clear the formation, and it would fire its missiles as it retired. In addition, P-500 Bazalt (SS-N-12 'Sandbox') and P-700 Granit (SS-N-19 'Shipwreck') missiles had a 'scout' feature in which one or more missiles in a salvo would try to identify the high-value target in a formation, cueing the others to attack it. At least in SS-N-19 that was done using a combination of radar and ESM. The 'scout' concept survives in the more recent P-800 Oniks/Yakhont (SS-N-26 'Strobile'), which in appearance is a scaled-down SS-N-19. In its case one of every three missiles turns on its radar. The 'scout' concept was also attractive as a way of reducing the radar and ESM signatures of the incoming missiles. SS-N-12 and, presumably, the later missiles turned on their radars at high altitude, estimated the course and speed of the target formation and then dived to low altitude, folding their radar antennas to reduce radar cross-section. They turned their radars back on when they crossed the estimated radar horizon of the target formation.

5. For example, the main air-search radar on board, say, the helicopter carrier *Moskva* would be distinguishable from the main air-search radar on board her sister *Leningrad*. This idea developed into the Specific Emitter Identification (SEI) technique currently used. It is the radar equivalent of identifying a gun by the pattern of grooves it impresses on a bullet, or by nicks in its firing pin.

6. This conclusion is contrary to that reached in the British Royal Navy Board of Enquiry into the Loss of HMS *Sheffield* in May 1982, which suggests the Exocet warhead did not detonate. One reason for the conclusion seems to be a lack of shrapnel damage. A copy of the report is located at: http://webarchive.nationalarchives.gov.uk/20121203133119/http://www.mod.uk/NR/rdonlyres/9D8947AC-D8DC-4BE7-8DCC-C9C623539BCF/0/boi_hms_sheffield.pdf

7. All opinions expressed are the author's own and should not be attributed to the US Navy or to any other organisation with which he has been associated.

4.3 TECHNOLOGICAL REVIEW

Author:
Jan Ziolo

SMERAS

(SUBMARINE ESCAPE, RESCUE, ABANDONMENT & SURVIVABILITY)

The crew on board a downed submarine (a DISSUB or distressed submarine in accordance with standard SMERAS terminology) face far more hazards than the crew of a distressed surface vessel. To take to a lifeboat, life-raft or to enter the water is a measure of last resort for any mariner. For a submariner, other dangers have to be faced and overcome even to reach this stage. The complexity of underwater search and rescue is clearly spelled out in the *International Aeronautical and Maritime Search And Rescue (IAMSAR) Manual*, the standard, approved quick-reference guide for any seafarer who may be required to assist in such a situation: '… medical care requirements for survivors of an underwater or submarine accident is specialized and competent medical advice is required …'[1] In fact, all aspects of underwater search and rescue are more complex than an equivalent situation on the surface.

The Royal Norwegian Navy's *Ula* class submarine *Utvaer* pictured operating on the surface off the Norwegian coast in May 2015. As can be imagined, the crew on board a downed submarine face far more hazards than the crew of a distressed surface vessel. *(Petter Brenni Gulbrandsen/Norwegian Armed Forces)*

These include vessel tracking, communications and alerting, search, casualty management and recovery, and – ultimately – the salvage of the casualty hull from the seabed.

Submarine forces around the world place strong emphasis on crew skills and training, the integrity of the vessel and its systems, and multiple layers of equipment redundancy. Prevention of a distressed submarine situation is infinitely better than cure. However, accidents and incidents still occur. The high levels of skill and motivation required of a submarine volunteer means that visible means of securing their survival need to be in place. Even for navies which still rely on conscription, the potential damage to reputation in losing a submarine and her crew would be considerable. For all of these reasons, a global, interlinked network of submarine rescue forces exists, which constantly strives to refine procedures and equipment. This is in response to the primary lesson learned from every submarine sinking: speed of response is king.

In this chapter, 'escape' refers to survivors exiting a DISSUB and ascending to the surface in either escape suits or escape bells. 'Rescue' refers to survivors being shuttled off the DISSUB by manned or remotely-operated rescue submersibles. 'Abandonment' refers to survivors exiting a DISSUB on the surface, and 'survivability' refers to any measure that maintains and extends a habitable environment on-board the DISSUB.

SURVIVABILITY – SELF-HELP MEASURES

Before looking at external search and rescue processes, it is worth noting that many submarine incidents are resolved by the internal resources of the boat and her crew. Submarines are constructed

All aspects of underwater search and rescue are more complex than the equivalent situation on the surface. Consequently thorough preparation and training – as well as effective international cooperation – are required to ensure the speedy and effective response needed to bring an incident to a satisfactory conclusion. The 'Black Carillon' exercise series allows Pacific Rim members of the International Submarine Escape and Rescue Liaison Organisation (ISMERLO) to test their submarine rescue capabilities. In this image, the RAN *Collins* class submarine *Farncomb* is acting as the exercise distressed submarine. The view is what either a submersible or – as in this case – the ROV pilot would see. Note that, for exercise purposes, the submarine has lights illuminated and high-visibility markings have been placed around the escape hatch to aid docking training for new personnel. *(Royal Australian Navy)*

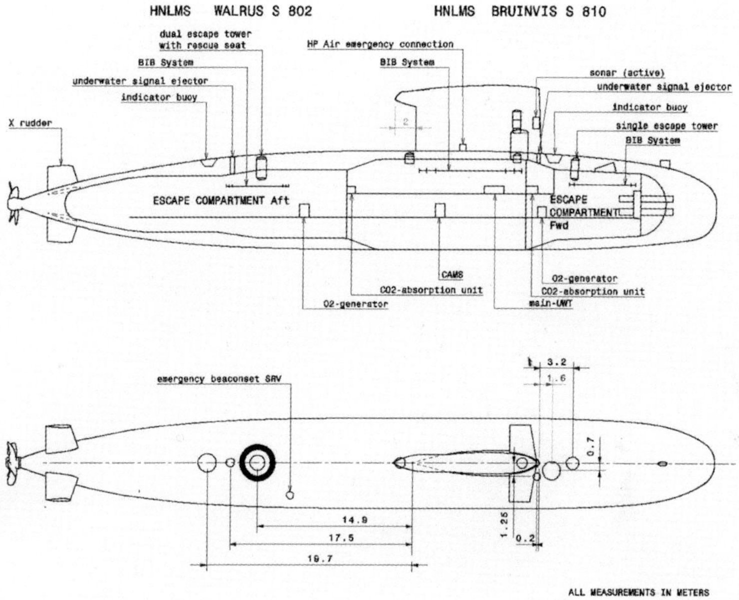

The standard submarine search and rescue reference for NATO navies – and the 'de facto' global standard – is the open source NATO Allied Tactical Publication 57 – *ATP 57: The NATO Submarine Search and Rescue Manual.* This includes technical rescue profiles for as many submarines as possible, including diagrams showing the position of escape hatches and other equipment. This is the diagram contained in the manual showing relevant details for the Dutch submarines *Walrus* and *Bruinvis. (NATO)*

and equipped to demanding standards, and have high levels of survivability built in to them. Submarine hulls are immensely strong constructs, faired-in for much of their surface area under an external 'casing' that acts as an effective 'crumple zone' in event of frontal collisions. Watertight integrity is paramount; every submariner should be able to isolate every bulkhead valve on their own submarine as a matter of routine. Multiple, high-capacity bilge and ballast pumps can be quickly lined up to pump out floodwater.

In the case of a fire, it should be noted that submarine firefighting systems make more use of high-expansion foam than seawater so as to avoid compromising a submarine's buoyancy or trim. This is especially the case forward and aft: areas that tend to house the torpedo spaces and propulsion machinery, both areas of high fire risk. Moreover, if the internal atmosphere is compromised by fire or another contaminant, many submarines are fitted with a ring-main of pressurised, diver-quality air – referred to in the Royal Navy as the 'Emergency Breathing System' (EBS) – that can be accessed via hundreds of quick-disconnect couplings using full-face masks as used by firefighters. This will allow the crew to function effectively, if not comfortably, to provide time for the boat to be surfaced or brought to periscope depth so as to allow ventilation of the hull to be carried out. However, sometimes external support will be needed, this is considered further below.

SUBMARINE TRACKING

In common with all search and rescue operations, the shorter the search phase, the greater the prospect of survivor recovery. To this end, peacetime submarine movements are closely controlled and monitored, in a manner similar to a scheduled airline flight. Operating water is allocated, along with pre-arranged diving and surfacing times and routine signal transmission times. In the event of several missed signals in turn – a submarine may not be able to come to a transmitting depth owing to surface traffic or sea state, or may have a defect or damage to her communications equipment – this narrows down the search area. IAMSAR procedures for a missing or overdue vessel define three phases, viz.

- **Uncertainty:** The vessel is overdue. The parallel SMERAS phase is SUBLOOK and the exercise equivalent codeword SMASHEX 1.
- **Alert:** The parallel SMERAS phase is SUBMISS and the exercise equivalent codeword SMASHEX 2.
- **Distress:** The parallel SMERAS phase is SUBSUNK and the exercise codeword SMASHEX 3.

It will be immediately apparent that a submarine deployed on operations – particularly if they are covert – will have less freedom to communicate regularly, surface or come shallow to 'snort' to either run auxiliary power or to clear out the internal atmosphere. As such, rescue forces will have less idea of the distressed submarine's position and may well not have access to conduct escape or rescue operations.[2]

The Royal Canadian Navy submarine *Chicoutimi* – the former Royal Navy *Upholder* – berthing at Faslane in Scotland in October 2004 following a fire at sea that resulted from water ingress whilst running on the surface. The fire resulted in the death of one of her crew as a result of smoke inhalation. Submarines' low freeboard means that down-flooding through open hatches is a risk that has to be guarded against. *(Canadian Armed Forces)*

SUBMARINE SEARCH AND RESCUE: GENERAL PRINCIPLES

A key principle of SMERAS is national primacy: the nation that has sustained the loss retains operational and tactical control of the search and rescue effort. In peacetime, the nation will declare a SUBMISS/SUBSUNK Phase through the International Submarine Escape and Rescue Liaison Organisation (ISMERLO). This maintains a protected website that allows global monitoring of unfolding situations as well as the status of declared submarine escape and rescue assets. A national submarine operating authority declaring a DISSUB situation can post updates and request assets as required.

The standard submarine search and rescue reference for NATO navies – and the 'de facto' global standard – is *NATO Allied Tactical Publication 57 (ATP 57)*, the NATO Submarine Search and Rescue Manual. This is an open-source document that sets out all the key procedures to be followed. It also maintains technical rescue profiles of as many submarines as possible, relying on navies and manufacturers making available unclassified rescue-specific data. This information typically includes internal volume, crew numbers, estimated crew endurance before air is exhausted, type and location of onboard escape stores, and external submarine plans showing positions of escape and access hatches, potential obstructions and salvage valves.

When dealing with a submarine search and rescue operation, the key initial questions to be answered by rescue forces are (i) where is the distressed submarine and (ii) in what depth of water does she lie? Modern rescue submersibles can effect a rescue from depths of up to 600m. Live, successful escape exercises have been conducted from depths of up to 200m.

It will be readily appreciated that most of the open ocean floor, the abyssal plains, rises, seamounts and trenches, all lie at far, far greater depths than this. Clearly, a distressed submarine that cannot maintain neutral buoyancy, establish positive buoyancy or drive to the surface is going to sink until the pressure hull implodes or a hull penetration fails, causing a catastrophic internal flood. However, a review of peacetime DISSUB incidents shows that

Royal Australian Navy submariners practice a surface abandonment exercise from the lead *Collins* class submarine. Surface abandonment offers much lower risks to survivors than other forms of escape and rescue. *(Royal Australian Navy)*

the bulk occur not over the abyssal plains but over the shallower coastal margins of the continental shelves – within the rescuable and escapable depths of water. These areas are also where surface vessels are concentrated and collisions are therefore more likely.

SURFACE ABANDONMENT

In event of an emergency, and assuming it is safe to do so, a submarine will aim to surface and run up auxiliary power generators and blowers to ventilate the internals and to recharge compressed air banks. This also allows crew to be protected from internal hazards – such as radiation, fire or atmospheric contaminants – by rotating personnel out of the pressure hull and onto the casing. Given that submarines have low freeboard when they are surfaced, there is a risk of down-flooding if hatches are opened in heavy seas. The near-loss of the Royal Canadian Navy's *Chicoutimi* (the former Royal Navy *Upholder*) on her delivery voyage in 2004 was a result of this. Inflatable freeboard extenders can be fitted in build or retrofitted to existing submarines' escape towers to mitigate this risk.

It may be, however, that the submarine is losing buoyancy or is otherwise untenable. In this case, the commanding officer may order a surface abandonment. Recent developments have provided specialist equipment to support this survival mode. For example, some submarine arms, such as the Royal Netherlands Navy have dispensed with indicator buoys and have used the available under-casing volume to install life-rafts that can be deployed either on the surface or when dived. Surface abandonment offers much lower levels of risk to survivors. It is becoming the preferred option in SMERAS.

SURVIVAL ON A DISTRESSED SUBMARINE

If, however, the distressed submarine is on the seabed, continued survival on board requires management of four overriding priorities. In descending order of importance, these are (i) protection, (ii) location, (iii) water, and (iv) food. Dealing with these is a complex task for distressed survivors. To simplify the process – and to aid effective, timely decision-making – a 'Guardbook' is included in on-board escape equipment. This is a sequenced process flowchart that is backed up by graphs, tables and survival instructions. The Guardbook defines what actions need to be taken in terms of making a compartment as safe as possible, how key parameters such as internal pressure, radiation and contaminants are to be monitored, how and when to release detection equipment such as pyrotechnic signals and indicator buoys, and when preparations need to be made for escape, if this is practicable

Protection: The survivors on board will need to try to arrest any flooding or leaks of pressurised air mains. These will drive up the internal boat pressure and, over time, saturate the bodily tissues of the survivors in a similar manner to saturation divers. Controlled decompression becomes more and more critical to survival and recovery as the duration of exposure to pressure increases. In addition, any fires, ruptures to pressurised gas systems such as nitrogen, or damage to weapons will contaminate the submarine's internal atmosphere. This could render it untenable for life, especially if the internal pressure is high. On a submarine carrying any nuclear materials, monitoring to ensure that containment systems are intact will also be required to ensure that survivors do not start to accrue a radiation dose.

To minimise oxygen consumption and carbon dioxide build-up, physical activities will be reduced as much as possible. Artificial means to keep both of these parameters within safe levels will be established from escape stores. The distressed submarine will rapidly become a cold and damp environment as its temperature equalises with the surrounding sea and condensation builds up on all surfaces.

Location: Getting the location of the distressed submarine known by others is often correlated with the time delay between the last position report and the incident. Once on the bottom, survivors may be able to release buoyant communication devices, in addition to submarine smoke signals. Escape compartments are fitted with independently-powered underwater telephones that can be used to alert search forces equipped with sonar to the distressed submarine's location. During previous submarine accidents, survivors have improvised various means to attract the attention of rescuers. These have included releasing buoyant materials from the submarine, as well as hull taps.

Water: Escape compartments can be fitted with an emergency fresh water tank to sustain life. Survivors

Instrument mannequins dressed in Submarine Escape Immersion Equipment (SEIE) seen inside the engine-room escape trunk of the US Navy submarine *Virginia* (SSN-774) in August 2004 during a series of pre-commissioning trials. SEIE is designed to allow safe escape from a submarine up to depths of around 200m. Survivors on a distressed submarine will become progressively more debilitated over time and this can make escape more difficult. *(US Navy)*

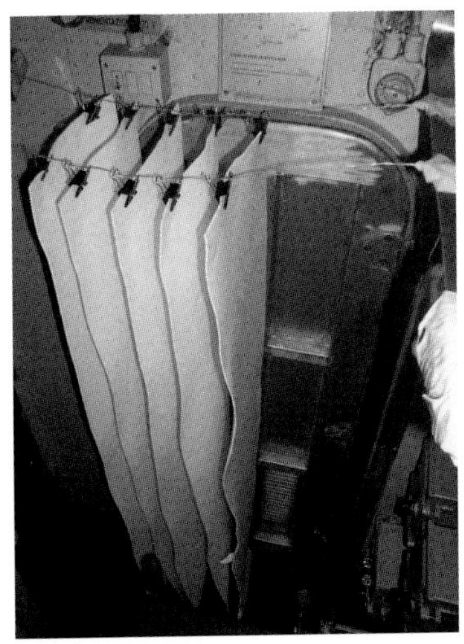

Artificial means – such as the use of carbon monoxide absorbent materials like these ExtendAir lithium hydroxide curtains manufactured by America's Micropore – can be used to help control carbon monoxide build-up on a distressed submarine. *(Micropore Inc.)*

A member of the Royal Navy's Submarine Parachute Assistance Group (SPAG) carries out a maritime parachute descent during an exercise over the Bay of Gibraltar in 2011. The SPAG can rapidly deploy a team with specialist escape and rescue knowledge to provide assistance to the crew of a distressed submarine; survivors on the surface and/or other forces involved in the rescue effort. The entry into service of the new A-400 Atlas transport aircraft – much larger than the current C-130 Hercules – is likely to expand the amount of equipment the SPAG can deploy. *(Crown Copyright 2011)*

The initial reaction to the occurrence of a distressed submarine will be carried out by rapidly deployable response assets, such as lightweight intervention remotely operated vehicles (ROVs) such as this Australian Scorpio 45. A British Scorpio, deployed half-way round the world by a C-17 Globemaster strategic transport, was responsible for the successful recovery of the trapped Russian *AS-28* submersible in 2005. *(Royal Australian Navy)*

will look to augment this by any means, from salvaging the submarine's internal fresh water reserves to collecting internal condensation.

Food: Escape food stores are similar to life-raft/lifeboat stores on surface vessels. They need to provide as much energy as possible, whilst absorbing as little water as possible while being digested.

It is to be noted that a prolonged stay in these conditions will leave survivors more and more debilitated, both physically and mentally. Ultimately a point will be reached when they are unable to function.

If escape from a distressed submarine needs to be supported, an escape gear ship will be equipped and dispatched to the distressed submarine's datum (location). Once on-scene, the ship will take over the co-ordination role from initial intervention units. The rescue ship needs to have sufficient deck space to accommodate all the relevant rescue equipment; an offshore platform support vessel will often be used. This image shows the Norwegian *Ram Star* carrying the NATO Submarine Rescue System (NSRS) during a 2011 training exercise. *(Torgeir Haugaard/ Norwegian Armed Forces)*

THE INITIAL REACTION

The initial reaction to the emergence of a distressed submarine will be carried out by response assets. These are small, lightweight and rapidly-deployable. They can therefore be on-scene quickly to establish communications with the distressed submarine, support an escape, and report as much information as possible on the situation back to the submarine operating authority. Response assets typically include air-portable remotely-operated vehicles, their operators and support equipment. These can be rapidly transferred to a vessel of opportunity (VOO).[3] VOO location and status is held within an ISMERLO VOO database to allow rapid identification, contact with the owners, and chartering should a distressed submarine situation so demand.

Once the vessel of opportunity and the equipment embarked reach the distressed submarine, they will first undertake an initial assessment of the situation. This will include establishing signs of life on board, the depth of water, the angle at which the distressed submarine is lying, and the condition of the escape hatches. The last-mentioned need to be able to make an effective seal with a rescue submersible in order to allow transfers of personnel or equipment. On occasion, a deployed remotely-operated vehicle can carry out additional actions. An example is the response to the underwater entrapment of the Russian *AS-28* 'Priz' class submersible in August 2005. In this case, the rapid deployment of the UK SCORPIO remotely operated vehicle team, by means of a Royal Air Force C-17 Globemaster strategic heavy-lift transporter, allowed the team to use the vehicle's manipulator arms to cut entangling sonar array cables away from the *AS-28*.[4] The submersible was then able to surface and its crew exit the vessel.

Several submarine operating authorities can also call on a 'Submarine Parachute Assistance Group' (SPAG) capability, either as a permanent asset declared to ISMERLO or as an ad-hoc capacity. SPAGs and their equipment can be deployed on to the ocean surface at the distressed submarine's datum (location) either by parachute or by a suitable helicopter. In accordance with IAMSAR procedures, the SPAG will assume the role of incident On-Scene Co-ordinator by establishing communications with onshore rescue authorities and the relevant submarine operating authority, taking over this role from the submarine's senior survivor

SPAG team members are trained to establish underwater communications with the survivors on board the distressed submarine and to assess the overall situation. They should have sufficient SMERAS expertise to provide guidance and support to both the senior survivor and the submarine operating authority; the latter to ensure that priority is given to the correct assets in the operation. The SPAG will be equipped with underwater telephones, satellite communications and VHF/UHF radio sets to assist effective communication. A key decision is whether to attempt an escape in the knowledge that shelter and specialist care awaits directly overhead, or to remain on the distressed submarine and await submersible rescue. Air-portable inflatable boats and life-rafts, diving medical supplies, food and water are all included within immediate reaction stores inventories, although the location and specifics of the incident will dictate what is drawn for use. The lightly-equipped SPAG will deploy when environmental conditions permit, but will require additional support from a larger responding vessel as soon as possible. Even with aerial resupply, conditions on board a heavily-loaded life-raft with debilitated survivors will be bad. The risks involved in deploying a SPAG must be weighed up against the benefits. The loss of many survivors from the British submarine *Truculent* in the River Thames in 1950, for example, provide a vivid illustration of a case where lack of on-scene shelter or support led to post-escape fatalities.[5]

An interior view of NSRS during a training exercise with the Royal Norwegian Navy. A crew member can be seen examining the connection between the submersible and the exercise submarine's escape hatch; the submersible's separate control area can be glimpsed through the hatch in the background. Use of a specialised submersible or rescue bell is the only effective response when a distressed submarine is lying in deep waters. *(Torgeir Haugaard/Norwegian Armed Forces)*

NSRS undergoing crane launch trials at RNAD Coulport during 2011. Jointly owned by France, Norway and the United Kingdom, it is regarded as one of the world's most advanced rescue submersibles and can operate to a depth of over 600m. Its survivor compartment can be mated directly to a recompression chamber once recovered, avoiding the worsening of barotrauma or decompression illness that might result from fleetingly exposing survivors to surface pressure. *(Crown Copyright 2011)*

The US Navy Pressurised Rescue Module (PRM) element of its Submarine Rescue Diving Recompression System (SRDRS) is recovered during an exercise in the Pacific. The PRM, based on the former innovative but troubled Australian *Remora* system, is driven remotely from the associated rescue gear ship. The system has replaced the pair of manned deep submergence rescue vehicles (DSRVs) of the *Mystic* class. *(US Navy)*

SUBSEQUENT INTERVENTION

If escape is a possibility, an escape gear ship will be equipped and dispatched to the distressed submarine datum. This will be loaded with stores and equipment, such as medical supplies, fresh clothing and bedding for survivors, and a portable recompression chamber. Additional medical and escape and technical personnel will also be embarked. Once on-scene, the escape gear ship will take over the role of on-scene co-ordinator and start the recovery, triage and treatment of survivors. Escapees are likely to be suffering from diving-related signs and symptoms, such as barotrauma and decompression illness, sustained during the rapid pressure increase in the escape tower and the equally rapid pressure decrease sustained during the free ascent to the surface. The likelihood and severity of these conditions will increase with escape depth. The escape gear ship's medical personnel will screen and triage escapees. If a recompression chamber is available on the EGS, priority for the limited spaces within will go to casualties who are most likely to respond positively to hyperbaric treatment. Escapees may present to the medical personnel with other injuries or conditions sustained during the initial downing of the submarine, such as the effects of smoke inhalation or burns. However, escapees are likely to be 'walking wounded' as more severe casualties will probably have been left in the distressed submarine.[6]

Rescue – transfer from the submarine to the surface by means of a specialised submersible or rescue bell – is the definitive solution to a distressed submarine situation and the only response when the vessel is lying at greater depths. The equipment required to conduct a rescue can be considerable and

The former Russian submarine *Juliett 484* breaks the surface of the waters of the Providence River, Rhode Island in June 2008 after a successful salvage operation led by US Army and Navy divers. The submarine – a Cold War relic turned into a museum ship – had sunk in around 10m of water. Most submarine salvage operations are considerably more complicated; often being contracted out to specialist firms. *(US Navy)*

CASE STUDY 1: BAP *PACOCHA*/*KIOWA MARU*, CALLAO ROADS, 1988

The Peruvian Navy's *Pacocha* was a US-built patrol submarine (the former *Atule*) that had been modernised under the 'Guppy' programme. She was running on the surface, in the vicinity of Callao, in August 1988 when involved in a night-time collision with a Japanese ice-strengthened fishing vessel, the *Kiowa Maru*. The *Pacocha*'s pressure hull was penetrated and the submarine rapidly sank, illustrating the vulnerability of a submarine running on the surface. The low profile and freeboard of a surfaced submarine do not lend themselves to ready identification by a tired and overworked officer of the watch or lookout, and collision damage can easily remove the small reserve of buoyancy that a surfaced submarine has available.

Of the forty-nine crew aboard *Pacocha*, the commanding officer and three others died in the immediate aftermath of the collision and sinking. Twenty-three were able to abandon the vessel, twenty of whom were subsequently recovered alive. This left twenty-two survivors mustered in the forward torpedo room. One officer's success in shutting the forward torpedo room hatch could not be attributed to anything other than a miracle, given its weight and the inrush of water; this miracle was later officially ratified by the Vatican!

The *Kiowa Maru* apparently made no official notification of the event – larger vessels such as the current generation of 200,000-tonne container ship or Very Large Crude Carrier may run over a submarine and not even notice the impact. In this case, the alarm was raised when the *Pacocha* became overdue for arrival. Radio communications between *Kiowa Maru* and her port agent, suggesting the possibility she had been involved in a collision, were intercepted and passed to the naval base. A tug was despatched to the scene in time to see a red distress flare from the downed submarine ignite on the water, illuminating the crew who had abandoned the submarine on the surface.

The Peruvian Navy had equipped all submarines with the Steinke Hood Escape Apparatus, but in common with the US Navy, had elected not to provide practical, 'wet' escape training.[1] A decision to remain on-board the distressed submarine pending rescue had to be altered as the atmosphere began to deteriorate rapidly.

US Navy Undersea Rescue Command personnel monitor a recompression chamber during a hyperbaric treatment. Note the CCTV, which along with audio communications, provides constant surveillance of both patients and inside attendants. Diving and hyperbaric-trained doctors are also required to provide continuous subject-matter expertise, and all personnel will be trained to at least IMCA Diver Medical Technician level. As demonstrated by the Peruvian *Pacocha* rescue, the likely number of patients, and potential severity of their condition mean that submarine survivor recompression is likely to be a different proposition to either saturation divers or recreational divers. Faster decompression tables and ruthless triage are likely to be applied. *(US Navy)*

Instructions on the operation and use of the Steinke Hood were passed to the survivors. Lacking confidence in the use of the equipment and unfamiliar with the associated hazards in escape, several crew members hesitated while exposed to increased escape depth pressure in the escape tower. This, coupled with steadily rising atmospheric pressures in the torpedo room, led to a range of decompression injuries and illness symptoms. Pressure-related 'barotrauma' was also sustained due to correct ascent drills not being followed. There was one further fatality at this stage.

LESSONS LEARNED

Pacocha was lost in close proximity to Callao, near to a concentration of rescue assets but also in congested waters. Her loss demonstrated the vulnerability of a hard-to-see submarine with a low margin of buoyancy in such conditions.

The rescue saw both surface abandonment and tower escape, the latter ably supported by diver intervention when they were able to re-seat the escape tower's jammed external hatch. The speed of the incident resulted in persons in 14° Centigrade water, lacking flotation or thermal protection. Close proximity to *Pacocha*'s homeport allowed rescue forces to be on-scene swiftly. There is some debate as to the utility of distress pyrotechnics in a modern age of satellite-based search and rescue; in the case of *Pacocha*, the red distress flare saved lives, as it provided positive confirmation of a SUBSUNK, the approximate datum of the DISSUB and clear indications of persons in the water.

The response also displayed commendable initiative on the part of rescue forces, as well as the need for regular training in the particulars of submarine escape and rescue. Small, cost-effective steps towards safety include effective pre-planning, notably ready access to submarine plans showing hatches and salvage fittings, as well as stockpiles of critical connectors and adaptors. Lack of crew confidence and familiarity in the escape equipment led to delays in escape, and the aforementioned barotrauma injury. The stresses inherent in making a submarine escape are not to be underestimated, and to maximise chances of survival, crews need as much confidence and experience as they can in the operation and use of the equipment. Local recompression facilities were rapidly overwhelmed by the large numbers of casualties. Time was lost tracking down naval divers owing to lack of means of communication and these in turn lacked specialist knowledge and had to learn on the job. Studies had to be made of one of the *Pacocha*'s sister submarines to note details of escape fittings, submarine hatches etc. Once on-scene, hull taps from inside were noted and replied to, but no agreed code was in place to allow effective communications. Messages had to be passed between survivors and rescuers via the submerged signal ejector, increasing the boat pressure with each operation. A decision was made to blow fresh air through the submarine via its salvage connections, but when this operation started, the survivors quickly shut the hull valve as – to their eyes – water present in the line was an indicator of flooding.

1 A Steinke hood – named after its inventor Lieutenant Harris Steinke – is an inflatable life jacket with a hood that completely encloses the wearer's head, trapping a bubble of breathing air. The Steinke hood takes up little space when stowed and is well-suited for tropical waters, but if water temperature and resultant survival times are a factor, then full-length escape suits are required.

CASE STUDY 2: K-141 *KURSK*

The Russian Navy's *Kursk* was a Project 949 (NATO reporting codename 'Oscar') nuclear-powered cruise missile submarine. In August 2000, with 118 crew on-board, she was a key participant in the Russian Northern Fleet's high-profile 'Summer X' exercise. One of the exercise's main evolutions was to have been a torpedo live-firing by the *Kursk*.

Kursk's outfit of weapons included Type 65 torpedoes fuelled by high-text peroxide (HTP), a highly volatile fuel that is explosive when it comes into contact with water. It is likely that – despite the best efforts of Russian submarine personnel – compromised build quality, maintenance routines and/or a degraded material state of equipment led to HTP coming into contact with seawater. Whatever the root cause, the effects were clear; an initial blast that flooded the forward part of the *Kursk*'s pressure hull and put the submarine on the bottom of the Barents Sea in about 110m of water. A second and much larger blast then took place. All the ship's company forward of the reactor compartment would have died instantly during the first blast. However, the sheer strength of modern submarines is proved by the survival of twenty-three personnel in the aft machinery spaces.

These men were utterly reliant on rescue, as there was no means of escape fitted to these compartments. The reaction of the Russian Navy in the crucial first hours after the sinking was slow, as they had become accustomed to delays in communications from submarines and did not make a link between reports of underwater explosions in the area and lack of signal traffic from *Kursk*. Once it became apparent that an incident had occurred, the Russian Navy initially relied on its own submarine rescue forces, 'Priz' class submersibles operated from *Mikhail Rudnitsky*, a converted former timber carrier pressed into naval service. *Mikhail Rudnitsky*'s main attribute was her heavy-lift crane system to launch and recover her submersibles: other than that, she was a standard merchant vessel lacking the dynamic-positioning capability to remain precisely on station. The submersibles found themselves operating at the limits of their battery endurance and despite truly heroic efforts in locating and taking imagery of the submarine, the rescue forces were unable to carry out any further actions.

By this time, external offers of assistance had been made and refused, and it was to be five days after the sinking before a commercial dive support vessel, the Norwegian *Normand Pioneer*, was able to negotiate manifold bureaucratic hurdles and arrive on-scene with the British SRS submersible *LR5*, a team of divers and a remotely-operated intervention vehicle. These determined that there were no signs of life on the submarine, and the operation now became a salvage instead of a rescue. Subsequent analysis of the bodies of the survivors in the aft compartments suggested that they had commenced survival routines and had burnt oxygen generators to keep the atmosphere within breathable limits. The pressure in the compartments had steadily risen to the point that the oxygen-rich air had been ignited. There has been debate as to how long the survivors endured. Certainly, *Kursk* lay in both an escapable and rescuable depth of water.

The Russian 'Priz' class submersible *AS-28* pictured surfacing in the Bering Sea in the afternoon of 7 August 2005 after a British Scorpio 45 remotely-operated vehicle had released it from entrapment by the aerial of a sonar array off the Kamchatka Peninsula. It appears lessons had been learned from the Russian Navy's slow response to the 2000 *Kursk* submarine disaster and a prompt international response meant that this incident had a much happier ending, with all seven crew members of *AS-28* surviving. *(US Navy)*

the designated rescue gear ship is likely to be an offshore platform support vessel with sufficient deck space to accommodate all necessary materiel.[7] Rescue units train regularly to ensure that they can rapidly mobilise, transport their equipment and fit it to a rescue gear ship. Included in this process is personnel call-out, calling for strategic airlift, setting up road haulage and clearing road routes, nominating suitable airports for departure and arrival, identifying suitable port facilities able to accommodate the rescue gear ship alongside and mate rescue equipment to the deck, and clearing all administrative hurdles along the way. This is clearly a complex process and there is a constant risk of delay.

Given this, additional intervention capabilities can be used to extend the survival time onboard a distressed submarine until rescue can be effected Emergency life support stores are particularly important. Large watertight and pressure-tight pods with critical items such as oxygen generators, carbon dioxide removal equipment, and water and medical supplies will be loaded on the rescue gear ship. These are transferred to the distressed submarine by a remotely-operated vehicle. The distressed submarine's survivors will rig an internal pod posting bag inside the escape tower, shut the inner hatch, flood the tower, equalise its pressure and open the outer hatch. The pod will then be loaded into the bag using the remotely-operated vehicle's manipulator arm, and the tower then shut and internally vented to allow the survivors to receive the stores. Every time this operation is carried out, the internal pressure in the distressed submarine will increase.[8] Other internal leaks are also going to contribute to this; for example the crew who had survived the initial explosions on board the Russian submarine *Kursk* were in this situation, as the propulsion shaft seal leaked when the shaft was stationary. To this end, some ISMERLO members hold emergency ventilation and decompression systems within their intervention inventories. Connected to a distressed submarine's hull valves using remotely-operated vehicles or a diver in an atmospheric diving suit (ADS), these can provide fresh, clean air as well as allowing controlled decompression *in situ*.

Once rescue commences, triaged personnel will be evacuated in batches for treatment at the surface. Recompression therapy may be required. The SMERAS world has begun to adopt the commercial saturation diving system of transfer under pressure

(TUP). Submersibles such as the NATO Submarine Rescue System (NSRS) can have its survivor compartment mated directly to the recompression chamber once recovered, preventing worsening of symptoms by fleetingly exposing survivors to surface atmospheric pressure. Other lines of technological development include remotely-operated rescue submersibles. The Australians led the way with their *Remora* project and this has now been taken further and matured by the US Navy. They have replaced their long-serving crewed deep submergence rescue vehicles (DSRVs) with the remotely-operated pressurised rescue module element of the Submarine Rescue Diving Recompression System (SRDRS). Medical treatment of survivors is likely to involve ruthless application of triage, as the high number of casualties is likely to swamp limited resources.

SALVAGE

There is a general presumption in favour of salvage of a distressed submarine's hull and the material contained within it where this is practicable, in order to minimise environmental impacts from weapons, fuel, lubricants and other potentially hazardous materials on board. The submarine's national operating authority will lead on the recovery, although this is likely to be subcontracted out to specialist survey and salvage companies with the requisite skills and equipment. The state of the submarine's hull will dictate recovery methods: most submarines are fitted with external hull 'salvage blow' valves that will allow internal floodwater to be pumped out, and compressed air to be pumped in. Other factors, such as how many other hull penetrations have been compromised, whether the boat is deeply embedded in sediment, and the fine buoyancy control required to safely surface the hull and keep it stable throughout the lift all play a vital role. It may be that the trauma of the initial event has fatally compromised the hull, perhaps due to an internal explosion, an excursion below hull crush depth, or from the collision with the seabed.

CONCLUSION: COMMUNICATION AND TEAMWORK

A submarine rescue effort is a complex effort where many specialist disciplines have to come together and deliver against the clock, in the full glare of global media. ISMERLO has led the way in pooling the actual and potential SMERAS assets of many countries, and then exercising them in accordance with a regular cycle. The capabilities and limitations of SMERAS assets also remain firmly within the OSINT or Open-Source Intelligence sphere, aiding understanding and rapid decision-making when the occasion arises. In peacetime, the submarine community will aim to work together to assist their fellow submariners. This mind-set needs to be fostered on a continuous basis.

SMERAS specialists often have to work hard to overcome a fatalistic perception that in event of a distressed submarine situation arising, there are never any survivors. There have been enough examples in the 125 years of submarine operations to make SMERAS a wise investment in event of a SUBSUNK situation. One of those could even be evolving as the reader completes reading this article.

The Royal Australian Navy submarine *Farncomb* departing Sydney with the Australian Defence Vessel *Ocean Shield* – a former offshore support vessel serving as rescue gear ship for the purpose of the exercise – at the start of 'Black Carillon' 2013. There have been enough examples of successful rescues from distressed submarines to make investment in SMERAS techniques a wise one. *(Royal Australian Navy)*

Notes

1. See the regularly updated *IAMSAR Manual - Volume 3* (London: IMO Publishing).

2. For security reasons, this chapter focuses on a scenario taking place during an overt operation, such as training within territorial waters.

3. Vessels of opportunity are merchant vessels, often drawn from the offshore energy sector, that have SMERAS-specific capabilities such as dynamic positioning, and sufficient deck strength and area to embark SMERAS equipment.

4. Strategic heavy transport aircraft, such as the C-5 Galaxy, C-17, A-400 Atlas and Antonov AN-125/225, are crucial elements of the ISMERLO declared asset database.

5. *Truculent* was struck by the Swedish tanker *Divina* whilst returning on the surface from post-refit sea trials. Most of the crew managed to escape the sunken vessel but the absence of support when they reached the surface meant many were swept away and drowned.

6. The senior survivor, when conducting escape tower runs, has to consider the risk of an escapee internally blocking the tower and rendering it unusable by those left inside the distressed submarine.

7. For example, the tripartite NATO Submarine Rescue System (NSRS) includes a submersible, a submersible portable launch and recovery system (PLARS), a flyaway recompression chamber, an intervention remotely operated vehicle (IROV), emergency life support stores and a rigid-hulled inflatable boat (RHIB) on an integral davit and cradle.

8. During an escape, those at the back of the queue to escape are far more likely to sustain a 'bend' owing to this effect.

Contributors

Edward Feege is Senior Maritime Intelligence Analyst at Gryphon Technologies LC. He has more than twenty years' experience in maritime intelligence, strategic communications and reporting in support of a wide range of US government, maritime and other customers. He is a graduate of the US Naval Academy, and during his naval career performed flight, maintenance, safety and operations duties for the airborne early warning community and was assistant division officer for the USN aviation manpower branch in the Office of the Chief of Naval Operations. His further qualifications include an MA in National Security Studies, Georgetown University, and MBA, George Mason University.

David Hobbs is an author and naval historian with an international reputation. He has written eighteen books, the latest of which is *The British Carrier Strike Fleet after 1945*, and has contributed to many more. He writes for several journals and magazines and in 2005 won the award for the Aerospace Journalist of the Year, Best Defence Submission, in Paris. He also won the essay prize awarded by the Navy League of Australia in 2008. He lectures on naval subjects worldwide and has been on radio and television in several countries. He served in the Royal Navy for thirty-three years and retired with the rank of Commander. He is qualified as both a fixed and rotary wing pilot and for eight years was the Curator of the Fleet Air Arm Museum at Yeovilton.

Theodore Hughes-Riley is a professional research scientist living in Nottingham, England. He obtained his bachelor's degree in physics from Lancaster University in 2009, and proceeded to graduate with his PhD from the University of Nottingham in 2014. His writing experience has typically been articles for scientific journals focussing on magnetic resonance and constructed wetlands. He has a long-standing interest in military matters, with a particular focus on maritime issues enthused by visits to Portsmouth and Chatham as a child.

Mrityunjoy Mazumdar has been writing on naval matters since 1999. His words and pictures have appeared in many naval and aircraft publications including *Jane's Defence Weekly*, *Jane's Navy International*, *Naval Forces*, *Ships of the World*, the USNI's *Proceedings* and *Warship Technology*. He is also a regular contributor to the major naval annuals. Mr Mazumdar lives in Vallejo, California with his wife.

Norman Friedman is one of the best-known naval analysts and historians in the US and the author of over forty books. He has written on broad issues of modern military interest, including an award-winning history of the Cold War, whilst in the field of warship development his greatest sustained achievement is probably an eight-volume series on the design of different US warship types. The holder of a PhD in theoretical physics from Columbia, Dr Friedman is a regular guest commentator on television and lectures on professional defence issues. He is a resident of New York.

Søren Nørby is a naval historian employed at the Royal Danish Defense College in Copenhagen. He has written extensively about the Danish Navy during the 20th and 21st centuries, including on submarines, fast attack craft, corvettes, frigates, Danish naval heroes and, most recently, the flexible support ships *Absalon* and *Esbern Snare*. He is currently working on a PhD about the transformation of the Danish Air Force between 1989 and 2011. Mr Nørby lives in Copenhagen; for more information on his work visit www.noerby.net

Tomohiko Tada was engaged in the development and design, manufacture, quality control and project management of defence equipment for the JMSDF at a major electronics company for more than thirty years. Following retirement, he has studied the history, current status and future trends of defence technology. He is currently a defence technology researcher and writer for the monthly magazines *Japan Military Review*, *Ships of the World*, *MARU* and the weekly magazine *World Weapons*. He is author of The *World's Warship Catalogue* and *Shipboard Weapons of the World* and co-author of the *United States Navy Handbook*, *The World's Near Future Weapons*, *21st Century Weapon Technology: Naval Ships and Aircraft*, the *Review of Today's European Navies*, the *Japan Maritime Self Defence Force Destroyer Perfect Guide*. He is a commentator on television and lectures on defence issues.

Guy Toremans is a Belgian-based, maritime freelance correspondent. His reports, ship profiles and interviews are published in the English language naval magazines *Jane's Navy International*, *Naval Forces* and *Warships IFR*, as well as in the French *Marines & Forces Navales* and the Japanese *J-Ships*. Since 1990, he has regularly embarked on NATO, Asian, South African and Pacific-based warships.

Scott Truver is Director, Team Blue, at Gryphon Technologies LC, specialising in national and homeland security, and naval and maritime strategies, programmes and operations. Since 1972 Dr Truver has participated in numerous studies and assessments – most notably supporting the inter-agency task force drafting the US National Strategy for Maritime Security (2005) – and has also written extensively for US and foreign publications. He has lectured at the US Naval Academy, Naval War College and Naval Postgraduate School, among other venues. His further qualifications include a PhD in Marine Policy Studies and a MA in Political Science/International Relations from the University of Delaware.

Jan Ziolo works in the North Sea oil and gas sector. He has been a professional marine engineer since graduation from UMIST in 1997 with a degree in Mechanical Engineering. Since then, he has worked at sea in both the Merchant Navy and the Royal Navy, where he earned his submariner's 'dolphins' and served as a Nuclear Engineer Officer of the Watch on *Vanguard* class submarines. He also served as a Staff Officer and Maritime Operations and Tactics Instructor at the Britannia Royal Naval College, as an Escape and Rescue Instructor at the Submarine Escape Training Tank, Gosport, and led the UK Submarine Parachute Assistance Group. Jan is also active as a powerboat skipper and instructor, having also served as a volunteer lifeboat crewman with the Royal National Lifeboat Institution, and as a recreational scuba diving instructor. He is a Chartered Marine Engineer with the Institute of Marine Engineering Science and Technology.